Introduction to plant population ecology

Second edition

Introduction to plant population ecology

Second edition

Jonathan W. Silvertown
Lecturer in Biology, The Open University

Copublished in the United States with
John Wiley & Sons, Inc., New York

Longman Scientific & Technical,
Longman Group UK Limited,
Longman House, Burnt Mill, Harlow,
Essex CM20 2JE, England
and Associated Companies throughout the world.

Copublished in the United States with
John Wiley & Sons, Inc., 605 Third Avenue, New York, NY 10158

First published 1982
Reprinted 1984
Second edition 1987

British Library Cataloguing in Publication Data

Silvertown, Jonathan W.
Introduction to plant population ecology. – 2nd ed.
1. Plant populations
I. Title
581.5'248 QK911
ISBN 0-582-44701-1

Library of Congress Cataloging-in-Publication Data

Silvertown, Jonathan W.
Introduction to plant population ecology.

Bibliography: p.
Includes index.
1. Plant populations. 2. Vegetation dynamics.
3. Botany – Ecology. I. Title.
QK910.S54 1987 581.5248 86–27835
ISBN 0–470–20708–6 (USA only)

Set in 10/12 Times Roman

Produced by Longman Singapore Publishers (Pte) Ltd.
Printed in Singapore.

Contents

Contents

Author's acknowledgements 1st edition

As with most books, the name on the cover belongs to only one of the people responsible for the final product. I would like to thank Eva, Alfred, Adrian and the friends who made the task easier by their tolerance of my awkward presence or my awkward absence during its preparation.

A number of colleagues and friends read parts of the manuscript and saved me from blunders. Though I have not accepted all the advice offered, I am very grateful to John Barkham, Steve Bostock, Brian Charlesworth, Deborah Charlesworth, Richard Croucher, Alastair Ewing, Mike Fenner, John Harper, Dave Kelly, Steve Newman, Pat Murphy, Steve Prince, Deborah Rabinowitz, Irene Ridge and Francis Wilkin.

Beverley Simon typed the manuscript with her usual skill and good humour and Niki Koenig drew the plant silhouettes with imagination and care. The staff of the Open University Library relentlessly pursued obscure references and kept me well supplied with scientific papers. I am also grateful to Dennis Baker and to the staff of Longman who guided the book through to publication.

Jonathan W. Silvertown
Open University

This book is dedicated to all students fighting racism

Author's acknowledgements 2nd edition

This is a complete revision of the first edition, but it is based upon that book and I again thank all those people mentioned on p. viii who helped me in writing it. For your support, Donna, and once more Adrian, Alfred and Eva – thank you.

I am grateful to the many people who sent me solicited or unsolicited comments on the first edition and to those others who sent me unpublished manuscripts. I hope their trouble is rewarded in the changes and additions I have made. I thank Viggo Andreasen, Alan Berkowitz, John Connolly, Phil Dixon, Rhonda Janke, Richard Law, Chuck Mohler, Steve Newman, Sunny Power, Deborah Rabinowitz and Heather Robertson who read parts of the manuscript, and in particular I thank Sana Gardescu and Ben Plumpton who both read nearly all at very short notice.

I appreciate the help of my colleagues at the Open University Irene Ridge, Peggy Varley, Alastair Ewing and Phil Parker who made it possible for me to take sabbatical leave at Cornell University where this book was completed. I am grateful to the Section of Ecology and Systematics at Cornell, to the 1985 students of Plant Ecology BS463, and most especially to Deborah Rabinowitz.

Jonathan W. Silvertown
Open University

Publisher's acknowledgements

We are grateful to the following for permission to reproduce copyright material:

Academic Press Inc (London) Ltd and the authors for fig 7.7 from fig 1 (Lovett Doust 1980), fig 1.3 from fig 1 (Penalosa 1983), fig 4.16 from figs 1c & d, 2a & d (Willey & Heath 1969); Acta Forestalia Fennica for fig 6.4 from fig 12 (Oinonen 1969); Bell & Hyman Ltd for fig 7.13 from fig 3 (Salisbury 1942); Blackwell Scientific Publications Ltd for fig 8.7 from fig 4 (Aarssen & Turkington 1985a), figs 6.3b & a from figs 3 & 4 (Hartnett & Bazzaz 1985a), fig 2.5 from fig 1b (Thompson 1986), figs 7.18a–c from fig 5 (Venable & Levin 1985), fig 8.8 from fig 1 (Watt 1974); Blackwell Scientific Publications Ltd and the authors for fig 4.11 from fig 6a, fig 4.12 from fig 1 (Ford 1975), fig 7.14 from fig 1.5 (Grime & Jeffrey 1965), fig 2.12 from fig 1 (Harper et al 1965), fig 7.6 from left hand side fig 7 (Harper & Ogden 1970), fig 4.10b from fig 2a (Kays & Harper 1974), fig 4.10a from fig 4a (Kays & Harper 1974), fig 6.2a from fig 1, fig 6.2b from fig 2 (Langer, Ryle & Jewiss 1964), fig 2.13 from fig 3 (Mellanby 1968), fig 4.13 from fig 4 (Mohler, Marks & Sprugel 1978), fig 2.8 from figs 7, 8 & 9 (Sarukhán 1974), fig 5.9 from figs 8, 16 & 20 (Sarukhán & Harper 1973), fig 2.6a & b from figs 6a & 7c (Thompson & Grime 1979), fig 6.5 from fig 1 (Turkington & Harper 1979), fig 4.1b from fig 8 (Watkinson & Harper 1978); Blackwell Scientific Publications and the University of California Press for fig 4.9 from fig 2.8 Copyright 1980 by Blackwell Scientific Publications reprinted by permission of the University of California Press (White 1980); the Botanical Society of Japan for fig 5.17 from fig 2 (Kanzaki 1984), fig 5.18a from fig 12 (Kohyama & Fujita 1981), fig 5.16c from fig 1 (Naka & Yoneda 1984); Butterworth & Co Ltd for fig 3.4 from left hand side fig 14.3 (Fridrikson 1975); Dr Faille for fig 5.15b from fig 6 (Faille et al 1984); Duke University Press & Professor Joan M. Hett for fig 4.1a from fig 1 Copyright 1971 the Ecological Society of America (Hett 1971); the Ecological Society of America for figs 5.16a & b from fig 1 (Brokaw 1985) copyright © 1985 by the Ecological Society of America, figs 7.3a & b from figs 1 & 3 (Schemske 1978) copyright © 1978 by the Ecological Society of America; Ekologia Polska for fig 2.10 from figs 7a & b (Symonides 1977), fig 5.5 from fig 8.9 (1) (Symonides 1979a); the International Council for Research in Agroforestry, Nairobi, for fig 8.3

from ch. 2 (Michon 1983); Professor J. L. Harper for fig 2.7 from fig 4.1 (Harper 1977); Macmillan Journals Ltd and the author for fig 3.2b from part of fig 2 (Bennet 1983) copyright © 1983 Macmillan Journals Ltd; the National Research Council of Canada and the authors for fig 7.1 from fig 3 (Eis et al 1965); Nature Conservancy Council and author for fig 9.2 from fig 3 (Clymo & Reddaway 1972); Dr J. Ogden for fig 7.8 (Harper 1977 after Ogden 1968); OIKOS for fig 5.12 from fig 7 (Crisp & Lange 1976), figs 5.8a & b from quadrats 1–13 of fig 13 (Kawano et al 1982); The Open University for fig 7.5 from fig 26 Open University Unit 2 S364 © 1981 The Open University; PUDOC for fig 1.1 adapted from fig 1 (Harper & White 1971); Dr A. M. Schaffer for fig 7.12 from fig 22.2 (Schaffer & Schaffer 1977); Springer-Verlag Heidelberg and the authors for fig 1.2e from fig 2 (Bergmann 1978), fig 4.15 from figs 1c & d (Burdon et al 1984), fig 9.3 from fig 8 (Hobbs & Mooney 1985), fig 4.6 adapted from figs 4, 5a–c & 6a–c (Keddy 1982); University of Chicago Press for fig 7.10 from fig 4 (Leverich & Levin 1979) copyright by the University of Chicago Press; University of Colorado for our fig 7.4 from fig 3 (Law, Bradshaw & Putnam), fig 2.3 from fig 3 (Marchand & Roach 1980); Verlag Eugen Ulmer for figs 1.2c & f from figs 155 & 149 (Ellenberg 1978).

1
Introduction

Population ecology of plants

Ecology, broadly defined, is the study of interactions between species and their environment. Population ecology is a specialized branch of ecology dealing with the numerical impact of these interactions on a specific set of individuals which occur within a defined geographical area: in other words a *population*. A population ecologist is interested in the numbers of a particular plant or animal to be found in an area and how and why population sizes change (or remain constant). Hence information on the age distribution of plants, the fate of seeds and seedlings and on the predators which affect these potential members of the next generation is vital.

Although it has nineteenth-century antecedents, population ecology, and plant population ecology in particular, is a twentieth-century science. The nineteenth century was the golden age of taxonomy when botanists in Europe and North America spent years of painstaking work classifying plants and naming species. Specimens collected throughout the British Empire were sent to Britain for study. Taxonomists were engaged in a kind of giant stocktake of the biological resources of the British colonies. Kew Gardens was an important centre for this work (Brockway 1980). Via Kew rubber (*Hevea brasiliensis*) was introduced into South-East Asia from South America. Holland, France and Britain all introduced coffee (*Coffea arabica*) into their colonies. Like rubber, the main centres for coffee cultivation are now in non-indigenous areas (Purseglove 1968).

Plant ecologists working at the beginning of the twentieth century were strongly influenced by the nineteenth-century preoccupation with classification. Hence a good deal of time was spent classifying and naming plant communities, almost as if they were organisms in their own right. For a while an argument even raged about whether plant communities developed through fixed sequences of species, analogous to the development of individual organisms through a sequence of embryonic stages (Clements 1916; Tobey 1981). Although the integral view of plant communities was challenged soon after it emerged (Gleason 1926, 1927), the emphasis on community classification has remained strong in plant ecology into the 1970s.

During this period applied ecologists studying forests or field crops saw things differently. They were (and are) interested in the *number* of seeds that must be sown to obtain a crop with an economic *yield*; the *density* at which seeds should be sown; the effects of varying *proportions* in mixtures of species; the causes of mortality and the *quantitative effects* of competitors (weeds), predators (insect pests, etc.) and disease. These studies form the historical foundations of plant population ecology.

Plant demography and population dynamics

Stages in the plant life cycle provide useful intervals at which to analyse changes that take place in plant population size with time. Demography is the study of these population changes and their causes throughout the life cycle. The basic stages of a typical plant life cycle are illustrated diagrammatically in Fig. 1.1. This diagram is self-explanatory, but there is some conventional terminology applied to various stages depicted in it.

The seed population in the soil is generally referred to as a *seed bank* or a *seed pool*; the latter term is used in this book. The interface between the seed pool and the establishment of seedlings is often envisaged as an *environmental sieve*. Some seeds pass through it to successful seedling emergence, others die or remain dormant in the soil.

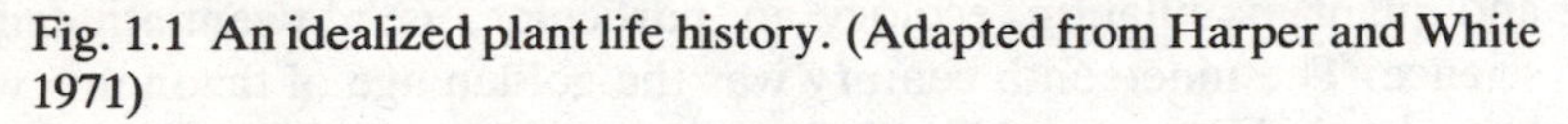

Fig. 1.1 An idealized plant life history. (Adapted from Harper and White 1971)

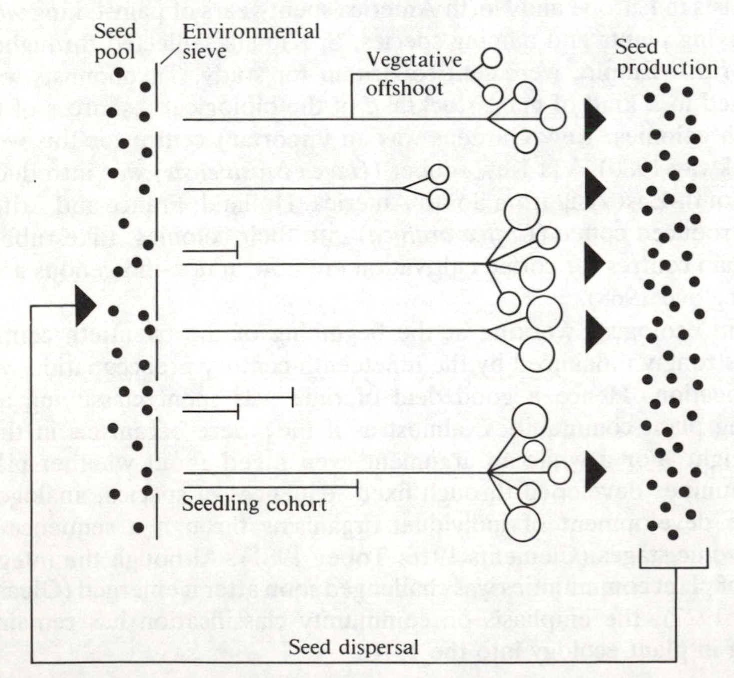

Seedlings which emerge simultaneously or nearly so form a *cohort*. The cohort is a particularly important unit in demography. By following the fate of a cohort of individuals through time, we can obtain values of probability for giving birth (seed production), death and survival for typical individuals of specific age.

The transition from juvenile stages such as seeds or seedlings to later stages in the life cycle in which reproduction occurs is called *recruitment*. Recruitment may occur from seeds or, as is the case in many plants, by the production of *vegetative* offshoots. These offshoots are initially physically attached to the parent, which may itself be no more than a collection of rosettes, stems or tillers (in grasses) each capable of an independent existence if they are detached. These morphological units with the potential for an independent existence are called *ramets*. Vegetative offshoots or ramets which have all been produced from the same parent constitute a *clone*. A plant, of whatever size and however divided into ramets, which originates from a seed is called a *genet*, as all parts share exactly the same genes.

Botanical custom is to divide species into annuals, biennials and perennials which notionally live for 1 year, 2 years and more than two years respectively. These terms are not precise enough for our present purpose. Some populations of 'annual' species actually live longer than 1 year (e.g. annual meadow grass). Though most biennials (e.g. carrot (*Daucus carota*), teasel (*Dipsacus fullonum*)) only flower once, many individuals take longer than 2 years to do so. One of these, hoary whitlow grass (*Draba incana*), is almost entirely perennial in Upper Teesdale, in Northern England (Bradshaw and Doody 1978b). A similar spectrum of life history is found in animals and plants and the same body of evolutionary theory applies to both, so we will adopt a universally applicable terminology.

Semelparous organisms are those which reproduce once and die (referred to in botanical nomenclature as monocarpic), *iteroparous* (or polycarpic) organisms reproduce more than once. We will use a restricted definition of the botanical terms *annual* and *perennial* to refer to organisms which live for 1 year and more than 1 year, irrespective of how often they reproduce.

Only four basic demographic processes determine how the total numbers in a population change in time: birth (B), death (D), immigration (I), and emigration (E). We can describe how these processes change the size (N) of a population between one time interval (t) and the next ($t+1$) with the equation:

$$N_{t+1} = N_t + B - D + I - E \qquad [1.1]$$

All statements about population dynamics (i.e. change) ultimately come back to this simple equation which is fundamental to many aspects of the ecology of plants. Figure 1.2 illustrates some of these aspects for Norway spruce (*Picea abies*) in Europe.

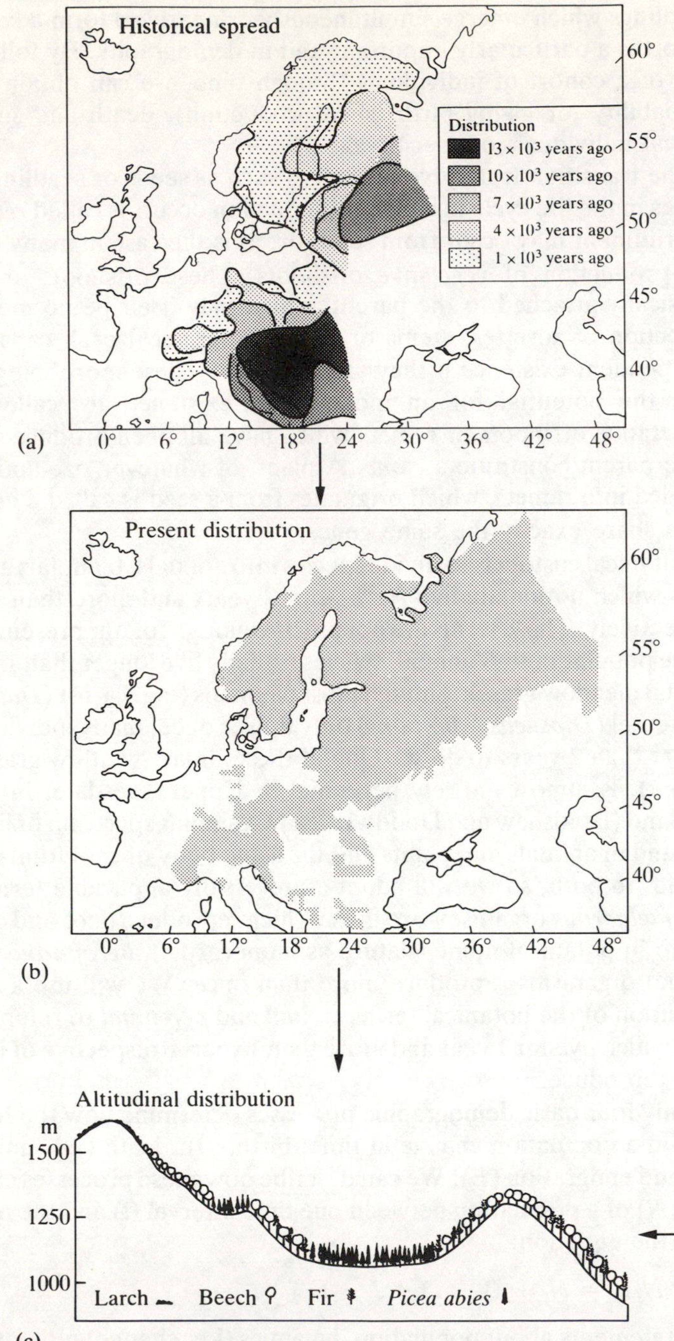
Historical spread
Distribution
13×10^3 years ago
10×10^3 years ago
7×10^3 years ago
4×10^3 years ago
1×10^3 years ago
60°
55°
50°
45°
40°
(a)
0° 6° 12° 18° 24° 30° 36° 42° 48°
Present distribution
60°
55°
50°
45°
40°
(b)
0° 6° 12° 18° 24° 30° 36° 42° 48°
Altitudinal distribution
m
1500
1250
1000
Larch
Beech
Fir
Picea abies
(c)

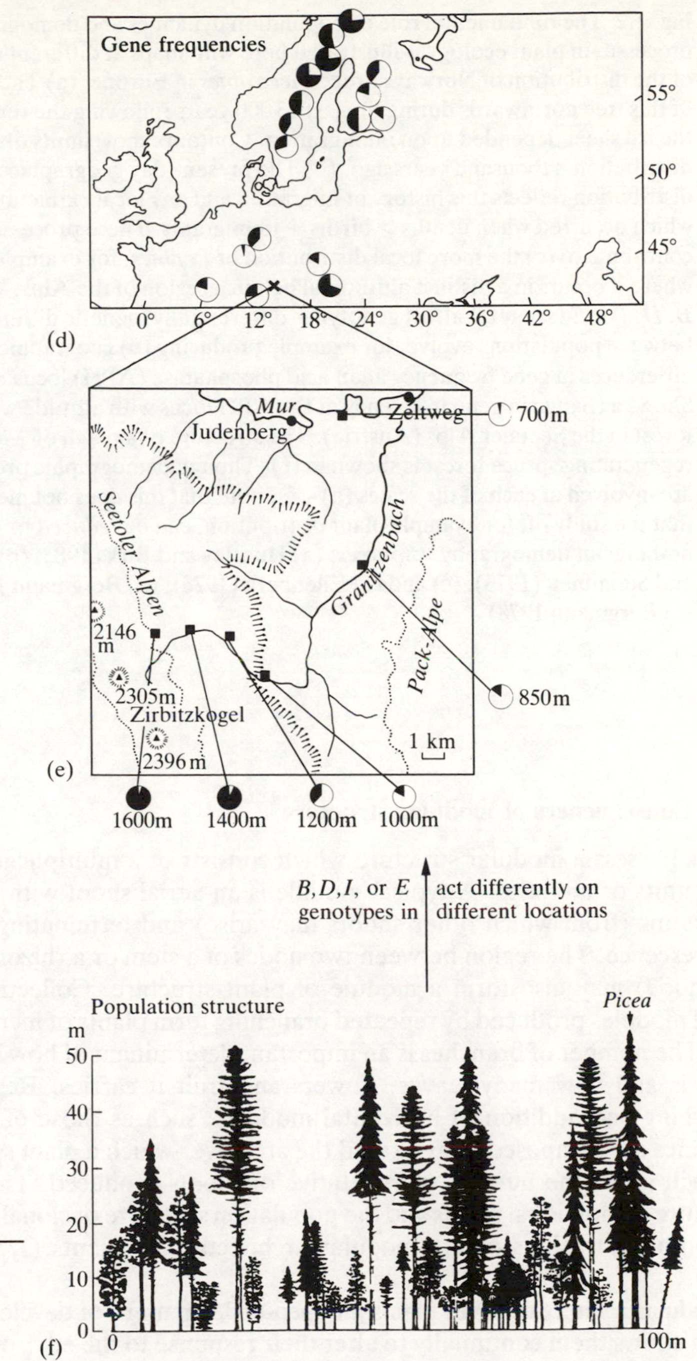

B, D, I, or E
act differently
n different
locations

Fig. 1.2 The fundamental role of population dynamics and demographic processes in plant ecology is illustrated here with maps at different scales of the distribution of Norway spruce *Picea abies* in Europe. (a) The spread of this tree northwards during the last 13 000 years following the retreat of the ice sheet depended upon immigration. Contours show limits of distribution *n* thousand years ago. (b) The present-day geographical distribution reflects this history of migration and any local extinctions which occurred when deaths > births + immigrants. These processes also contribute to (c) the more local distribution of *P. abies*, for example where it occurs in a distinct altitudinal belt in a region of the Alps. When *B*, *D*, *I* or *E* (see text) affect genotypes differentially, genetic differences between populations evolve, for example producing (d) geographical differences in gene frequency at an acid phosphatase (APH) locus. (e) Shows a change in gene frequency at the APH locus with altitude within a forest in the Seetaler Alps (Austria). The structure of a patch of regenerating spruce forest is shown in (f). Though demographic processes are involved at each of the scales (a)–(f), note that this does not mean that the study of, for example plant distribution, can be *reduced* to nothing but demography. (*Sources*: (a) Huntley and Birks 1983; (b) Jalas and Suominen (1973); (c) and (f) Ellenberg (1978); (d) Bergmann 1975; (e) Bergmann 1978)

Consequences of modular structure

Plants possess a modular structure which consists of a multiplication of basic units or modules. A typical module is an aerial shoot with lateral meristems (from which other shoots may arise) and terminating in an inflorescence. The region between two nodes of a stem or a rhizome (an internode) may also form a module of plant structure. Collections of aerial modules produced by repeated branching form plants of increasing size. The number of branches is an important determinant of how large a plant is and how many leaves, flowers and fruit it carries. Repeated branching and addition of horizontal modules, such as those of which rhizomes are composed, may extend the area over which a plant spreads and will affect the number of vegetative offshoots produced. Thus the structure of individual plants and the population structure of clonal plants are both expressions of the modular architecture of plants (J. White 1979).

Modular achitecture gives plants an open-ended pattern of development which allows them continually to alter their response to the environment by changing the direction of growth or type of plant organ produced. The structure of the tropical rainforest liana *Ipomoea phillomega* illustrates how modular growth permits plants to respond to the environment in ways which animals accomplish by behaviour (Fig. 1.3). This liana can produce several types of shoot, depending upon the nature of the light environment. In shaded conditions, stolons with long internodes and rudimentary leaves extend rapidly over the ground. When a gap is reached twining shoots with large leaves are produced and ascend to the tree canopy where the liana forms a crown of its own (Peñalosa 1983). The parallel with the behaviour of a foraging animal is difficult to resist!

Fig. 1.3 A map of the shoot system of *Ipomoea phillomega* on the floor of a tropical rainforest in Veracruz, Mexico. This plant originated at the 'manifold' which has an ascending shoot and a crown in the canopy. Liana crowns and ascending shoots are represented by circles. Stolons that have lost their tips end in a 'T' and those which are still growing are shown with a 'Y'. (Peñalosa 1983)

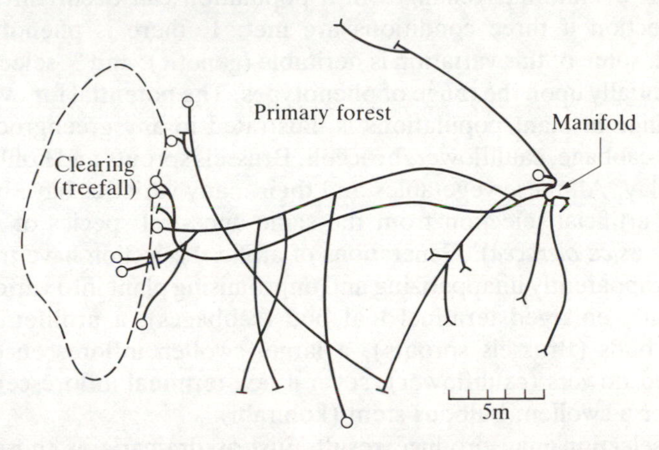

Population dynamics and evolution by natural selection

Most of this book is concerned with changes in the numbers of plants and with changes in the relative proportions of different species in populations. At a finer level of analysis we would find that most single-species populations consist of a collection of individuals which differ to some degree in characteristics such as leaf shape, flower colour, biochemical properties or other aspects of their outward appearance or *phenotype*. Differences in genetic make-up (*genotype*) often underlie phenotypic differences beween individuals. Only the phenotype is visible to us but this is the 'public face' of the genotype, some of whose properties may be deduced by growing different individuals in the same environment in transplant experiments, from breeding experiments with different phenotypes, or by electrophoresis which was the technique used to map gene frequencies in *Picea abies* (Fig. 1.2 (d & e)).

In populations of white clover (*Trifolium repens*) two distinct types of plant are found. When leaves are damaged, one type of plant produces free cyanide which is thought to deter slugs and other mollusc predators (Crawford–Sidebotham 1972). The other type of plant does not produce the cyanide poison. Thus there is a cyanogenic and a non-cyanogenic phenotype.

Breeding experiments have shown that two genetic loci determine cyanogenesis. One locus controls the production of cyanogenic glucosides and the other controls the production of an enzyme which breaks down the glucosides to liberate cyanide. The cyanogenic phenotype occurs only when the genotype has the correct combination of alleles for production of both the enzyme and the glucoside. There are three genotypes which have acyanogenic phenotypes: 1. the genotype lacking the glucoside allele; 2. the one lacking the enzyme allele; and 3. the genotype lacking both.

Significant evolutionary changes in a population can occur through natural selection if three conditions are met: 1. there is phenotypic variation; 2. some of this variation is heritable (genetic); and 3. selection acts differentially upon the range of phenotypes. The potential for evolutionary change in plant populations is illustrated in any greengrocer's shop where cabbage, cauliflower, broccoli, Brussels sprouts and kohlrabi are on display. All these vegetables and their many varieties have been derived by artificial selection from the same ancestral species of wild cabbage (*Brassica oleracea*). Generations of artificial selection have transformed this apparently unappetizing and unpromising plant into varieties with a greatly enlarged terminal leaf bud (cabbages), a proliferation of axillary buds (Brussels sprouts), a large, swollen inflorescence of undeveloped flowers (cauliflower), several lax, terminal inflorescences (broccoli) or a swollen, bulbous stem (kohlrabi).

Natural selection may produce results just as dramatic as those of artificial selection but is less rapid. It occurs when one phenotype leaves more descendants than another because of its superior ability to survive, or to produce offspring, or because of superiority in both of these characters. Notice that *survival* and *reproduction* are both demographic processes, and hence natural selection is also a demographic process. Where it is possible to analyse the demography of different phenotypes in a population separately, it is also possible to determine which phenotype is likely to leave the most descendants and hence in which direction(s) natural selection is acting.

Survival and reproductive success are combined into a single measure of relative evolutionary advantage called *fitness*. The fitness of a particular phenotype is not a fixed value, but is determined in the context of prevailing ecological conditions and relative to the survival and reproductive success of other phenotypes which occur in the same population.

For example, cyanogenic *Trifolium* may have a higher fitness than acyanogenic *Trifolium* in the presence of slugs but when slugs are absent the relative values of fitness for the two phenotypes may be reversed. Cyanogenic plants are more prone to frost damage than acyanogenic plants, and without slugs to redress this disadvantage or in particularly cold areas, the latter phenotype may survive and reproduce the more successfully.

Although population dynamics lays the basis for an assessment of selective forces in natural populations, this absorbing subject is mostly beyond the scope of this book. However, the relevance of natural selection in plant populations is briefly touched upon in most chapters, particularly in Chapter 7. In Chapter 2 we look at the four basic demographic processes and how they are analysed in more detail.

Summary

Population ecology is principally concerned with the processes which determine population size and population changes. Historically, this approach owes a great deal to the work of applied ecologists. Demography is the quantitative study of population changes throughout the life cycle. The phases of a model plant life cycle are the seed pool, the *environmental sieve*, the seedling *cohort* and various stages of juvenile development leading up to reproductive maturity. New plants may be *recruited* to the population from seed or from *vegetative offshoots* (or *ramets*), which have the potential for an independent existence. *Clonal* plants spread by the multiplication of ramets. A plant originating from a seed is a *genet*. The four basic demographic processes which determine changes in population size are *birth* (*B*), *death* (*D*), *immigration* (*I*) and *emigration* (*E*). Individual plants and clones both have a *modular structure* which confers an open-ended pattern of development that allows plants to respond to the environment by altered growth. This is analogous to the behavioural responses of animals.

Fitness is a relative measure of evolutionary advantage which is based upon the survival and reproductive success of individuals with different phenotypes. *Natural selection* is a demographic process.

2

Life tables and some of their components

A newspaper which specializes in titillating its Sunday readers with gossip and scandal proclaims on its masthead that 'All human life is here'. A somewhat naïve demographer, opening a paper with this masthead, would expect to find the contents crammed with life tables and fecundity schedules, for, as far as demography is concerned, in these all life may be found. Life tables and fecundity schedules summarize all the most important events in a population: the births, the deaths and – essential information to gossip writers and demographers alike – the *age* of the individuals who are dying or giving birth. Social scientists joke that demography is the study of people broken down by age and sex.

Life tables and fecundity schedules

Life tables were first drawn up by actuaries who needed a precise set of data on mortality in the human population in order to be able to assess the insurance risk that is attached to different individuals. Although people may die at any age and from a variety of causes, statistically (i.e. on average) the death risk to an individual is related to that individual's age. Life tables therefore divide the population into age classes, each of which has an *age-specific mortality risk*.

The simplest way of compiling a life table is to follow the fate of the individuals in a cohort from birth until the last member of the cohort dies. This 'following' procedure produces a *dynamic life table*. Another method, commonly used in situations where it is not practical to follow the demise of a cohort through time (e.g. for long-lived trees), estimates age-specific death risks from the age structure of a population at one moment in time. This produces a *static life table* and is discussed in Chapter 4.

A simplified example of a dynamic life table for an annual plant, *Phlox drummondii*, is shown in diagrammatic form in Fig. 2.1. Populations often contain overlapping generations, thus complicating the methods required to obtain a cohort of uniform age. This is particularly a problem in the seed fraction of plant populations, since dormant seeds often accumulate in the soil, year after year. Viable seeds of *P. drummondii* apparently do not persist in the soil beyond one season so no overlap of generations occurred in the seed fraction of this population. Furthermore, seeds germinated approximately synchronously, so that all members of the population effectively belonged to the same cohort and formed a *discrete generation*. Both of these factors simplified the collection of data considerably.

Fig. 2.1 Diagrammatic life table for *Phlox drummondii*. By convention, rectangles represent stages of the life-cycle, inverted triangles represent transition probabilities between stages and the diamond represents seed production.

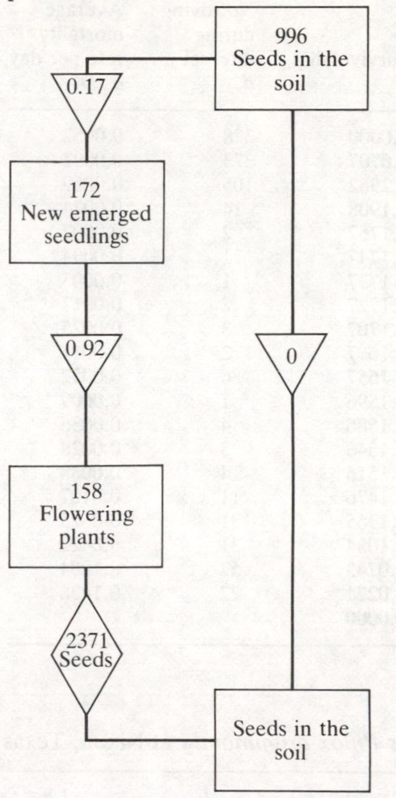

A census of the seed population was carried out seven times before germination and then at 7-day intervals until all remaining individuals flowered and died. All these data are shown tabulated in Table 2.1(a). The time intervals are indicated in the first column of the life table. The second column records the length, in days, between two successive censuses, the third column records the number of survivors N_x present at the beginning of an age interval and the fourth column the proportion l_x surviving to day x.

Various useful statistics can be derived from l_x values:

number dying during intervals: $d_x = (l_x - l_{x+1})\, N_x$

e.g. for $x = 0$: $d_x = (1 - 0.6707)\, 996 = 328$

proportion entering interval that do *not* survive: $q_x = l_x - l_{x+1}$

e.g. for $x = 0$: $q_x = 1 - 0.6707 = 0.3293$

average daily mortality rate $= q_x/D_x$

e.g. for $x = 0$: $q_x/D_x = 0.3293/63 = 0.0052$

Table 2.1(a) Life table for *Phlox drummondii* at Nixon, Texas

Age interval (days) $x - x'$	Length of interval (days) D_x	No. surviving to day x N_x	Survivorship l_x	No. dying during interval d_x	Average mortality rate per day q_x/D_x
0– 63	63	996	1.0000	328	0.0052
63–124	61	668	0.6707	373	0.0092
124–184	60	295	0.2962	105	0.0059
184–215	31	190	0.1908	14	0.0024
215–231	16	176	0.1767	2	0.0007
231–247	16	174	0.1747	1	0.0004
247–264	17	173	0.1737	1	0.0003
264–271	7	172	0.1727	2	0.0017
271–278	7	170	0.1707	3	0.0025
278–285	7	167	0.1677	2	0.0017
285–292	7	165	0.1657	6	0.0052
292–299	7	159	0.1596	1	0.0009
299–306	7	158	0.1586	4	0.0036
306–313	7	154	0.1546	3	0.0028
313–320	7	151	0.1516	4	0.0038
320–327	7	147	0.1476	11	0.0107
327–334	7	136	0.1365	31	0.0325
334–341	7	105	0.1054	31	0.0422
341–348	7	74	0.0743	52	0.1004
348–355	7	22	0.0221	22	0.1428
355–362	7	0	0.0000		

From Leverich and Levin 1979

Table 2.1(b) Fecundity schedule for *Phlox drummondii* at Nixon, Texas

$x - x'$	B_x^{seed}	N_x	b_x^{seed}	l_x	$l_x b_x$
0–299	0.000	996	0.0000	1.0000	0.0000
299–306	52.954	158	0.3394	0.1586	0.0532
306–313	122.630	154	0.7963	0.1546	0.1231
313–320	362.317	151	2.3995	0.1516	0.3638
320–327	457.077	147	3.1094	0.1476	0.4589
327–334	345.594	136	2.5411	0.1365	0.3470
334–341	331.659	105	3.1589	0.1054	0.3330
341–348	641.023	74	8.6625	0.0743	0.6436
348–355	94.760	22	4.3072	0.0221	0.0951
355–362	0.000	0	0.0000	0.0000	0.0000
					$\Sigma = 2.4177$

From Leverich and Levin 1979

Largely because actuaries were only interested in the relation between survival and age and had no use for information on births, fecundity data are traditionally recorded separately from the life table, but in a parallel fashion. The number of seeds produced by individuals (or females of dioecious species) in an age interval $x - x'$ is B_x, and age-specific fecundity per plant (b_x) is then B_x/N_x. This information and other measurements of seed production derived from it which we will use in later chapters are shown for *P. drummondii* in the fecundity schedule in Table 2.1(b).

In species where growth is determinate and the probabilities of survival and giving birth are closely related to age, it makes elementary sense to tabulate birth and death in terms of the age of individuals. Actuaries would lose a good deal of accuracy in their calculations if they based their estimate of someone's chances of survival on some parameter only loosely related to it such as body-weight. Age is not necessarily the best predictor of the fate of an individual in all species, and in particular in plants where growth is extremely plastic we may expect to find other parameters which are more useful.

Age versus stage

The practice of plant demographers in the Soviet Union since the 1940s has been to classify individuals in a population according to their *stage* of growth and reproduction rather than according to their chronological age (Gatsuk *et al.* 1980) T. A. Rabotnov and other Soviet ecologists have used this method in a large number of demographic studies, several of which have followed the fate of cohorts of herbaceous plants over periods of 10 years or more.

In one such study of the perennial *Ranunculus acris* growing in a floodplain meadow (Rabotnov 1964, 1978a) individuals in a 10 m^2 plot were classified into the categories juvenile (J), immature (I), vegetative (V), generative (flowering, G) or dead (D). When the first records were made in 1950 there were 178J, 125I, 123V and 25G. The fate of these plants was recorded the following year and is shown in Fig. 2.2(a). Nearly equal proportions of the 178 juveniles died or grew to the I stage, and small proportions remained J or became V. The 1950 cohort of immatures also had mixed fortunes, with as many plants remaining in the I stage as transferring to the V stage. Small proportions of the 1950 V's and G's actually regressed to earlier stages in 1951, but the majority in each cohort remained in the same stage of growth in which they were found during 1950. Only a very small proportion of the 1950 V's and G's died in 1951. It is clear from this study that population changes in *R. acris* are complex and that the fate of an individual plant is not strictly determined by its chronological age. In fact these kinds of transitions between one growth stage and another cannot be described in a life table which is designed for populations in which all individuals move inexorably from one age interval to the next.

Fig. 2.2 (a) The fate, 1 year later, of juvenile (J), immature (I), vegetative (V), generative (G) individuals of *Ranunculus acris* from a population marked in 1950; (b) the fate of survivors of the 1950 population between 1953 and 1954. Numbers in *italics* are the sample sizes for each growth stage. The width of arrows is proportional to the probability of the transitions between stages. (Data from Rabotnov 1978a)

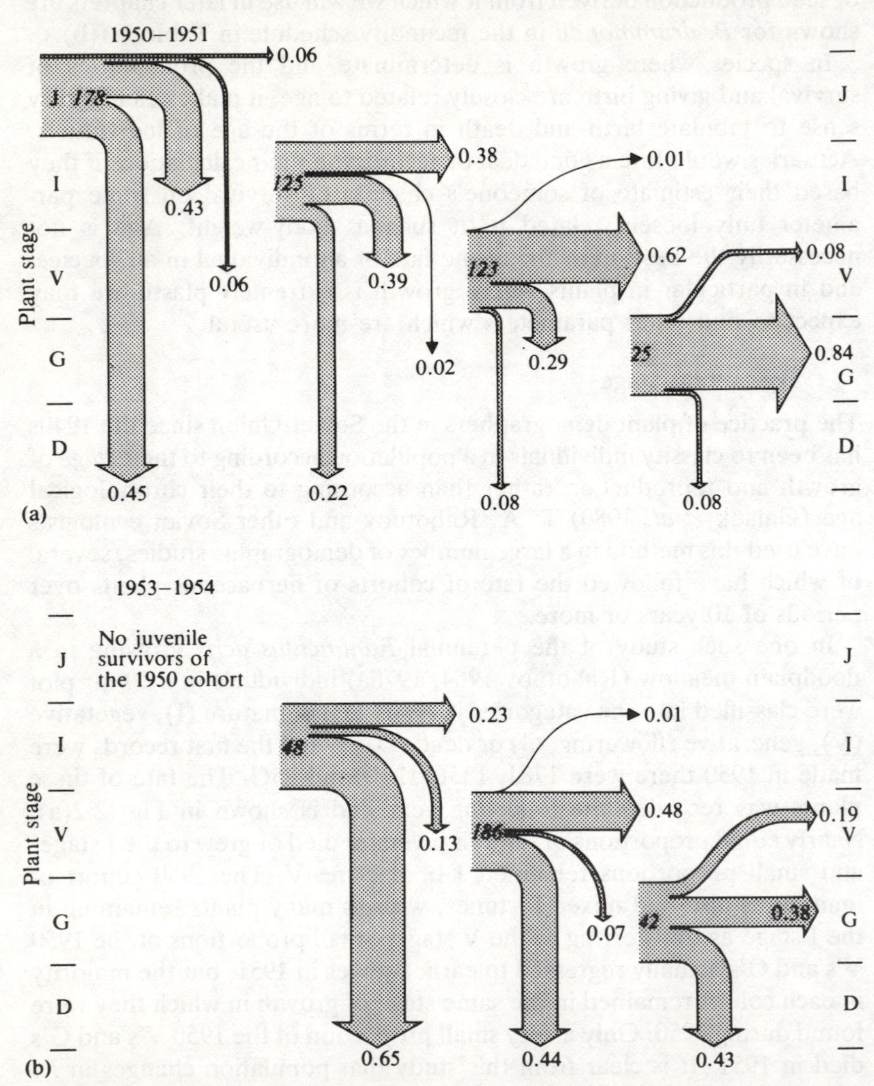

Plant populations appear to be governed by more probabilistic processes than those which determine the fate of individuals in the populations for which life tables were originally designed. Unfortunately we cannot place exact chronological ages on plants in the I, V or G stages to judge the importance of age in the *R. acris* study properly because some plants originated before 1950 when the study began. Nevertheless we know that *R. acris* does not escape the penalties of old age entirely because transition probabilities between growth stages did alter as the study progressed. Figure 2.2(b) shows the fate in 1953–54 of 276 survivors from the original plants of 1950. Death rates in all three remaining stages rose dramatically: from 0.22 to 0.65 among Is and from 0.08 to about 0.44 in Vs and Gs.

The relative importance of age versus stage of plant growth in determining the fate of individuals has been compared in experimental populations of teasel (*Dipsacus fullonum*) by Werner (1975) and Werner and Caswell (1977). This plant is generally referred to as biennial, but like many plants described in this way it actually often takes more than 2 years for an individual to reach flowering, which is always followed by death. Werner overcame the problem of determining the age of plants by sowing populations in field sites where *D. fullonum* was previously absent and following their fate from seedling emergence. Individuals were mapped over a 5-year period and rosette size and vegetative or flowering condition were noted. Rosettes were divided into different size classes and it was found that the probability of a rosette flowering was strongly correlated with its size but independent of its age. In contrast to the behaviour of *R. acris*, transition probabilities for rosettes at a given stage of growth (size) were appreciably the same whether the rosettes were 2, 3, or 4 years old.

Werner and Caswell (1977) compared the accuracy of predictions of population size made by matrix models (Ch. 3) based upon age-related transition probabilities and stage-related transition probabilities. They found that the stage-related models predicted changes in the number of seeds and vegetative or flowering rosettes found in experimental populations better than the age-related models, even though the transition probabilities for both types of model were derived from the same populations. The lower accuracy of the age-related models was evidently due to the fact that events in the life history of teasel are not as closely tied to age as they are to the growth stage of these plants.

Flowering behaviour is related to rosette size in most semelparous perennials including wild parsnip (*Pastinaca sativa*) (Baskin and Baskin 1979b), wild carrot (*Daucus carota*), common evening primrose (*Oenothera biennis*) (Gross 1981) and *Oe. erythrosepala* (Hirose and Kachi 1982) to name but a few of many that have been studied. The size of a rosette in these plants is usually an indication of the size of the tap-root beneath, in which carbohydrates are stored. The tap-root supplies these carbohydrates to the above-ground part of the plant when flowering occurs (Glier and Caruso 1973), so the size of the tap-root itself would probably provide a more accurate prediction of flowering. Unfortunately it cannot be measured easily without disturbing the plant.

The size of the storage organ is more easily measured in species which possess a bulb. The age and size of bulbs are closely related to each other in most species with such an organ. This makes the independent effects of age and size on flowering difficult to assess. Nevertheless flowering appears to be related to bulb size, independently of age, in commercial tulips (Fortainier 1973), and probably also in wild daffodil (*Narcissus pseudonarcissus*) (Barkham 1980) and a number of other bulbiferous woodland herbs (Kawano 1975; Kawano and Nagai 1975).

Dormant buds, oskars and life below ground

Analysing plant populations in terms of stages makes even more sense when we examine the recruitment phase of plant life histories. Plants exhibit a number of devices by which individuals persist for years in a dormant state awaiting a period of amelioration in the environment before they embark on full growth and the road to reproduction. The most common form of this waiting game is played by dormant seeds, but dormant buds on rhizomes play the same role in couch grass (*Elymus repens*) (Tripathi and Harper 1973) and sand sedge (*Carex arenaria*) (Noble, Bell and Harper 1979).

With some exceptions, trees do not possess dormant seeds in the soil. Instead, many tree seedlings may be found in populations of ageing juveniles, lingering in a stunted condition in the field layer, far beneath the tree canopy. We could describe this habit as the *Oskar* syndrome, after the character in Gunther Grass' novel *The Tin Drum*, who preferred the juvenile to the adult state and stopped growing at the age of three. Oskars occur in a variety of canopy species including hemlock (*Tsuga canadensis*), Norway spruce (*Picea abies*), white oak (*Quercus alba*), holly (*Ilex aquifolium*) and American beech (*Fagus grandifolia*) (Grime 1979) and many tropical species including dipterocarps in Malaysian rainforest and *Nectandra ambigens* in the neotropics.

The striped maple *Acer pensylvanicum* which occurs in the eastern USA has oskars on the forest floor where it occurs as an understorey tree. These oskars persist for up to 20 years suffering little mortality while they await an opening in the tree canopy. If an opening appears, they grow rapidly, flower and reproduce (Hibbs and Fischer 1979).

An analogous strategy is found in Prescott chervil (*Chaerophyllum prescottii*) which is a semelparous umbellifer found in meadows in the forest steppes of the USSR. This plant has no detectable reserve of dormant seed in the soil but possesses a dormant underground tuber which may vary from the size of a pea to that of a hen's egg. These tubers are produced from the transformed tap-roots of vegetative rosettes whose leaves and outer roots shrivel and die after several seasons' growth. The tuber left behind by this vegetative stage may remain dormant in the soil for over 10 years and it does not enter the next stage of growth until the surrounding vegetation is disturbed. When meadows are ploughed or areas of grass are killed under a haystack, hundreds of dormant chervil tubers are suddenly activated into growth. The tubers produce a system of surface roots, leaves are formed and flowering takes place, followed by death. Seeds germinate immediately they reach the soil, allowing seedlings to take advantage of the area of bare ground created by the same disturbance which originally stimulated flowering (Rabotnov 1964).

There are a number of examples of this kind of interrupted life history among terrestrial orchids. Orchid seeds are among the smallest produced by any angiosperm; they are so small that probably no autotrophic seedling could establish itself from such an impoverished beginning. The usual pattern of development of terrestrial orchids, following the germination of the seed, involves a more or less prolonged period during which the orchid depends upon a fungus for a supply of nutrients. During this parasitic phase of life, a cigar-shaped mycorrhizome, with no leaves or aerial parts, is formed. After subterranean growth of some years, the enlarged mycorrhizome produces an aerial shoot. Shortly afterwards the mycorrhizome itself disappears and is replaced by one or more tubers which contain no fungal hyphae. Underground development may be protracted, for instance the burnt orchid (*Orchis ustulata*) may take up to 15 years to produce its first aerial shoots and flowers. Thereafter, most orchids perennate by the production of new tubers and new shoots each year.

In some terrestrial orchids, plants which have already reached the flowering stage, and which have lost their mycorrhizome, may regress, developing a new mycorrhizome and returning to an exclusively underground existence. Red helleborine (*Cephalanthra rubra*), lady's slipper (*Cypripedium calceolus*) and creeping lady's tresses (*Goodyera repens*) are woodland species which behave in this manner when the tree canopy in their habitat closes, excluding light. Plants of red helleborine have been known to reappear after 20 years of subterranean life, when the tree canopy has opened again (Summerhayes 1968).

Certain vines of evergreen tropical rainforest exhibit seedling behaviour which incorporates both the Oskar syndrome and an underground storage organ. A small understorey shrub is produced which remains in a stunted condition for a considerable period of time. While in this state, the oskar produces a large tuber from its tap-root. Then, quite suddenly, the central stem of the shrub begins to elongate. Growing as much as 5 cm per day, it may reach more than 5 m into the tree canopy before it produces the leaves and clambering stems of a mature vine. Janzen, who describes the habit of these plants, suggests that the main advantage of this sudden switch from inactivity to rapid growth may be that it exposes the tender growing tip of the climbing stem to predators for as short a time as possible. A stem elongating more slowly and steadily would probably be more vulnerable. The tuber stores the resources which make a sprint for the tree canopy possible in these vines. The tuber itself is therefore potentially a rich, concentrated food source for predators. In fact it is less vulnerable than other parts of the plant because tubers are usually protected by large quantities of toxic secondary compounds (Janzen 1975a).

Seed dispersal

Plants are very poorly mobile by comparison with animals, for even sedentary animals such as barnacles have highly mobile larval stages. It is perhaps this very lack of migratory ability in plants which has reinforced the evolution of mechanisms which allow plants to persist through unfavourable periods in such a variety of ways.

Seeds are, of course, the main means of dispersal for higher plants and their movement is of interest to the population ecologist for two reasons: firstly, seeds may augment or deplete local populations, thus affecting population size; and secondly, small numbers of dispersing seeds may act as founders of new populations which may grow to significant proportions within a few generations. Pollen, which can move much greater distances than seeds, is a significant transporter of genes between established populations but cannot produce either of the ecological results of seed movement. The effects of both seed and pollen movement on gene flow are beyond the scope of this book but are comprehensively reviewed by Levin and Kerster (1974) and Handel (1983).

Seeds dispersed by wind, whether aided by a flight appendage such as a wing or a pappus or not, generally move only short distances from the mother plant (Sheldon and Burrows 1973). Figure 2.3 shows the

Fig. 2.3 Frequency distributions of dispersed seeds for four arctic alpine herbs. (From Marchand and Roach 1980)

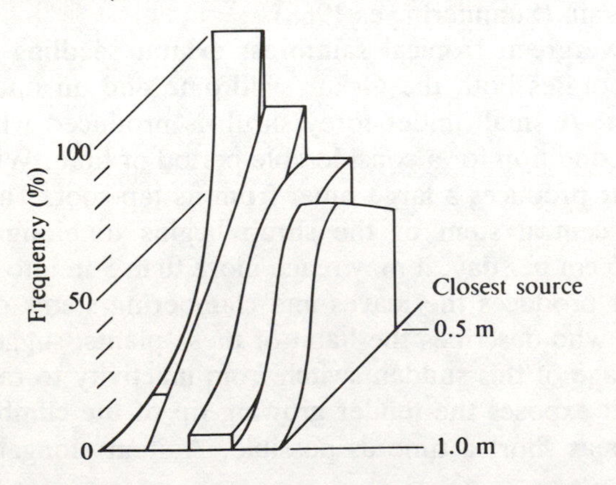

distribution of seeds with distance from plants of four species of arctic alpine herbs which colonize gaps in tundra vegetation. Although these plants are all colonists of ephemeral sites, their dispersal ability is not good. The shape of the dispersal curves for these species typifies patterns of seed fall found in a variety of other herbs, many of which deposit most of their seeds near the base of the plant, or at a short distance from it.

Seed dispersal by animals

Bats, birds, mammals, ants, earthworms and even tropical river fish (Gottsberger 1978; van Steenis 1981) disperse seeds and determine the distance seeds travel and the location and pattern of deposition by their behaviour. Animal dispersal is almost the rule in temperate woodland herb communities and for shrubs. It is common in trees of mesic forests in both the temperate zone and the tropics but rarer in lianas (Howe and Smallwood 1982; Gentry 1982). Plants with animal-dispersed seeds usually have fruit such as berries which attract dispersal agents or, like acorns and some other nuts, may be dispersed by animals that cache seeds or drop them while feeding.

Dispersal increases the distance between offspring and parent but not necessarily that between offspring themselves, which may be concentrated by the animal which deposits them. Figure 2.4 illustrates both long-distance transport and aggregation of offspring for a shrub with fleshy berries. However, it is not clear in this case how much aggregation is due to the direct effect of bird transport and how much to the patchy distribution of suitable sites for establishment. Probably both were involved.

The consequences of seed dispersal for later survival of the plant depend upon whether the main source of mortality is predation, which is often most severe around the parent where animals forage, or some other source such as pathogens which are more severe at high seedling density.

Rodents and birds cache seeds during periods of abundance but not all these may be recovered before they germinate. The pinyon pine (*Pinus edulis*) in North America and oaks (*Quercus robur*) in the Netherlands both depend on regeneration from seeds cached by jays (Ligon 1978; Bossema 1979). In Virginia, blue jays transported 54 per cent of acorns from a heavy oak crop and cached most under litter where they could germinate. Most of the rest were eaten by jays near parent trees or parasitized by insect larvae (Darley–Hill and Johnson 1981). Full-grown pines *Pinus flexilis* and *P. albicaulis* in Colorado and Wyoming, many of which appeared to have several trunks originating from a single root, turned out to be genetically separate individuals germinated from caches made by nutcrackers (Linhart and Tomback 1985).

Seed dispersal by ants is more than an incidental act in the course of

Fig. 2.4 The distribution of barberry *Berberis vulgaris* bushes (•) arising from seeds dispersed by birds feeding on fruits of an isolated barberry hedge (●●●●) in Fayette County, Iowa. (Stakman *et al.* 1927)

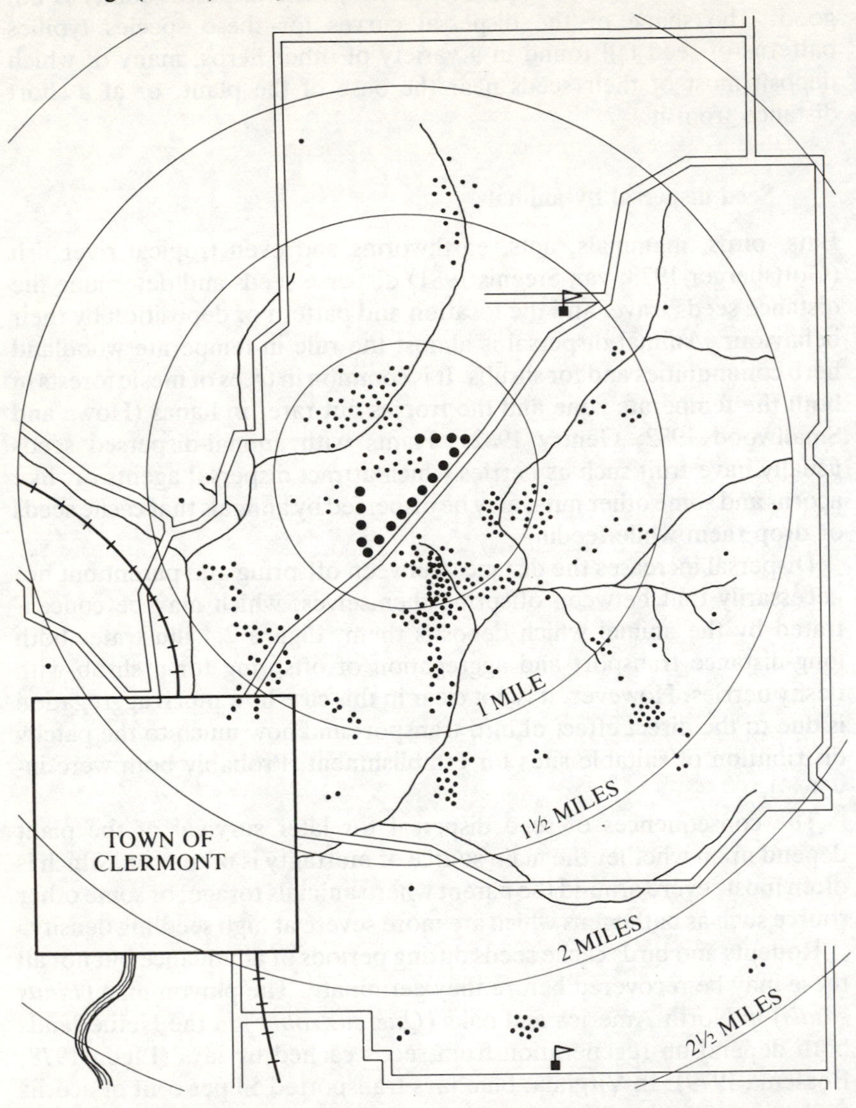

predation for certain plant species such as violets (*Viola* spp.). These have a special oil-containing appendage attached to the seed, known as an elaiosome, to which ants are attracted. Seeds with an elaiosome are carried off to the nest, the elaiosome is eaten and the seed itself, still viable, is deposited on the ants' refuse heap where it may germinate. In a study of ant/seed interactions (known as myrmecochory) in a group of violets growing in a forest in West Virginia, Culver and Beattie (1978) found that one species (*Viola papilionacea*) depended on ant transportation of its seeds to escape bird and rodent predators which ate any seeds not removed by ants. Most capsules of another species, *Viola nuttallii* dehisce in the morning between 9 am and 11 am when ants are most active in removing them and when granivorous rodents are least active (Turnbull and Culver 1983). Harvester ants appear to afford similar protection from rodents to *Datura discolor* in the Sonoran desert (O'Down and Hay 1980).

Seed dispersal of herbs by grazing animals can be important, even though the plants in question appear to have no especially attractive fruit. Janzen (1984) suggests that for large herbivores which consume plants with small seeds the 'foliage is the fruit' and there is good evidence for this in some habitats. Cow pats collected in a pasture newly cleared from the Amazon rainforest in Venezuela contained an average of 820 viable seeds of grasses and sedges in a concentration twenty times the density of the same groups found in the soil (Uhl and Clark 1983). Welch (1985) germinated seeds from the dung of cattle, red deer, sheep, rabbits and red grouse grazing heather moorland in Scotland. Cattle had the most copious loads: in the glasshouse one cow pat had 662 seedlings of 24 species germinate on it. Some plant species were found in the dung of some herbivores but not others as a consequence of their different diet selection.

Animal-dispersed seeds that pass through the gut of a herbivore require a thick integument to protect them. Alexandre (1978) found that fruit was an important part of the diet of elephants in a forest in the Ivory Coast and recovered seeds of 37 species of trees in their droppings. Many of these species have large fruit (5–10 cm) and do not appear to be dispersed by monkeys or birds. Trees with large fruit also occur in neotropical forests. Though large herbivores such as elephant-like gomphotheres once occurred in America, they have been extinct for 10 000 years. The extinction of these animals, which probably consumed some fruits, must have changed the composition of neotropical forests but in what way is very difficult to determine, particularly since other mammals may have taken over their role as dispersal agents (Janzen and Martin 1982; Howe 1985).

It is virtually unknown for any plant species to be dependent upon a particular animal species for dispersal to the extent that it would be endangered by the animals' extinction, but there may be one such plant and it is probably significant that it occurs on an oceanic island with a small fauna. The tambalacoque tree (*Calvaria major*) is endemic to the island of Mauritius in the Indian Ocean but by 1973 only 13 trees, all more than 300 years old, were known to exist in native forests (Temple 1979). This island was also the home of the dodo (*Raphus cucullatus*) whose eponymous extinction occurred before 1681. The nut of the tambalacoque is large, very hard and will not germinate without being abraded. The dodo fed upon fruits and seeds, had a large beak and a gizzard that contained large stones with which it crushed hard food. Fossil tambalacoque nuts have been found with dodo remains, making a convincing, if circumstantial, case for the possibility that this tree was dispersed by the dodo and that its nuts, armoured to survive the dodo gizzard, have been unable to germinate in the wild since the last dodo died.

Recruitment from buried seeds

A pool of buried seeds can be found beneath virtually all types of vegetation. Indeed, if we count the *genets* above and below 1 m^2 of almost any soil surface, many more individuals lie dormant below the surface than grow above it. Because seeds accumulate in the soil, the seed pool can act as a kind of 'memory', though a selective one, of previous vegetation at a site. Chancellor (1985) found that the distribution of weed seeds in an old arable field reflected three phases of past use. During one of these phases a fence had been erected across the site. The line of this old boundary was clearly visible thirty years later as a line of demarcation in the population of *Fumaria officinalis*, showing that this annual weed disperses very poorly and was still restricted to a distribution determined during the original infestation of the site.

Some typical seed population sizes for various types of vegetation are given in Table 2.2. Seeds are never uniformly distributed in the soil but are concentrated near the surface and are usually very patchy in abundance, even on a small scale (Fig. 2.5). This patchiness has almost always been overlooked in quantitative assessments of the seed pool and consequently nearly all studies have probably used too few samples to achieve a good estimate (Thompson 1986). The best sampling strategy for an

Fig. 2.5 The distribution of seeds of the grass *Danthonia decumbens* in 7 cm × 7 cm, 5-cm deep blocks of soil taken from a turf of acid grassland on Dartmoor, Devon, England. (Thompson 1986)

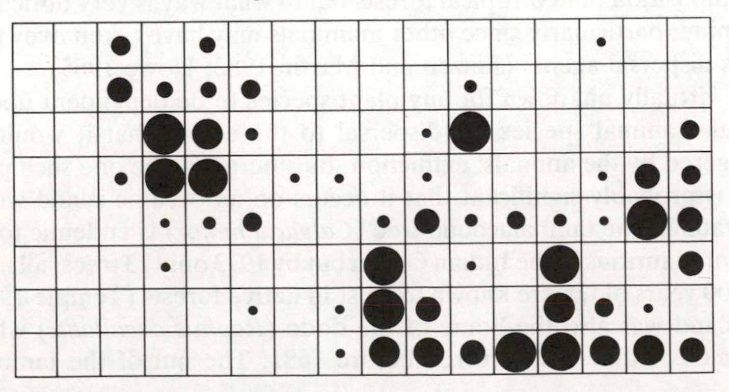

Table 2.2 Numbers of seeds and the predominant species present in the seed pools of various vegetation types

Vegetation type	Location	Seeds m^{-2}	Predominant species in the soil	Source
Tilled agricultural soils				
Arable fields	England	28 700–34 100	Weeds	Brenchley and Warrington 1933
Arable fields	Canada	5000–23 000	Weeds	Budd, Chepil and Doughty 1954
Arable fields	Minnesota, USA	1000–40 000	Weeds	Robinson 1949
Arable fields	Honduras	7620	Weeds	Kellman 1974b
Grassland, heath and marsh				
Freshwater marsh	N. Jersey, USA	6405–32 000	Annuals and perennials representative of the surface vegetation	Leck and Graveline 1979
Salt marsh	Wales	31–566	Sea rush where abundant in vegetation, grasses	Milton 1939
Calluna heath	Wales	17 500	*Calluna vulgaris*	Chippendale and Milton 1934
Perennial hay meadow	Wales	38 000	Dicotyledons	Chippendale and Milton 1934
Meadow steppe (perennial)	USSR	18 875–19 625	Subsidiary species of the vegetation	Golubeva 1962
Perennial pasture	England	2000–17 000	Annuals and species of the vegetation	Champness and Morris 1948
Prairie grassland	Kansas, USA	300–800	Subsidiary species of the vegetation, many annuals	Lippert and Hopkins 1950
Zoysia grassland	Japan	1980	*Zoysia japonica*	Hayashi and Numata 1971
Miscanthus grassland	Japan	18 780	*Miscanthus sinensis*	Hayashi and Numata 1971
Annual grassland	California, USA	9000–54 000	Annual grasses	Major and Pyott 1966
Pasture in cleared forest	Venezuela	1250	Grasses and dicot weeds	Uhl and Clark 1983
Forests				
Picea abies (100 yr old)	USSR	1200–5000	All earlier successional spp.	Karpov 1960
Secondary forest	N. Carolina, USA	1200–13 200	Arable weeds and spp. of early succession	Oosting and Humphries 1940
Primary subalpine conifer forest	Colorado, USA	3–53	Herbs	Whipple 1978
Subarctic pine/birch forest	Canada	0	No viable seeds present	Johnson 1975
Coniferous forest	Canada	1000	Alder *Alnus rubra*	Kellman 1970
Primary conifer forest	Canada	206	Shrubs and herbs	Kellman 1974a
Primary tropical forest	Thailand	40–182	Pioneer trees and shrubs	Cheke *et al.* 1979
Primary tropical forest	Venezuela	180–200	Pioneer trees and shrubs	Uhl and Clark 1983
Primary tropical forest	Costa Rica	742	Pioneer trees and shrubs	Putz 1983

accurate estimate of a patchy seed distribution is to use many small samples rather than a few large ones (Roberts 1981).

Though seed populations are of some interest in themselves, ultimately they are only of importance from the point of view of the plant when seeds are recruited from the seed pool to the growing population. In the light of this it is interesting to compare the species composition of seed pools with that of the active population.

In the majority of cases when buried seeds are identified and counted it is found that there is little direct relation between the abundance of a species above ground and the abundance of its seeds in the soil (Table 2.2 and Fig. 2.6(a)). Forest soils in late successional woodland predominantly contain the seeds of plants of earlier stages in the succession, and perennial grassland soils are replete with the seeds of short-lived species and contain few seeds of the dominant grass species. The exceptions to these discrepancies between the growing flora and the species occurring in the seed pool occur in arable fields (Fig. 2.6(b)) and in the seed pools of annual grassland such as those studied by Major and Pyott (1966) in California.

Fig. 2.6 The relative abundance of mature plants (histograms) and of seeds in the soil (vertical bars) for the major species of (a) a deciduous woodland and (b) an arable field. (From Thompson and Grime 1979)

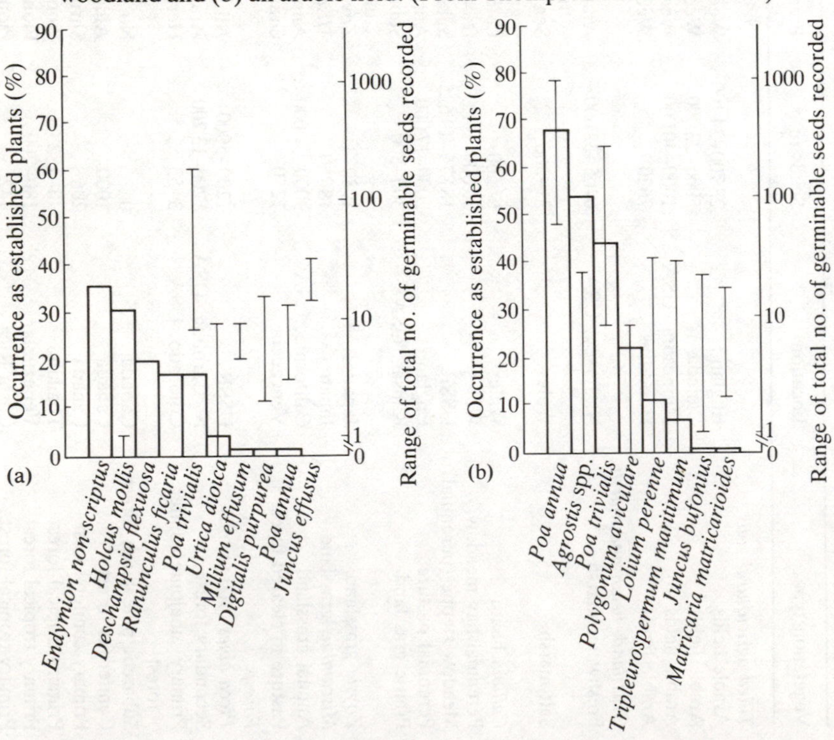

These exceptions suggest an explanation for the paradoxical composition of most seed pools. It is that dormant seeds are only produced in large numbers by species whose growing populations are subject to periodical local extinction. This is plainly the case for early successional plants, grassland annuals and arable weeds, all of which are usually well represented in the soil of their respective habitats.

Though the small seeds of short-lived plants dominate seed pools beneath many types of vegetation it is incorrect to assume that all short-lived plants rely upon buried dormant seed for recruitment. *Vulpia fasiculata* is a small annual grass of coastal sand dunes which possesses no seed dormancy. Seeds of this species germinate soon after they ripen and very few apparently become incorporated into a seed pool (Watkinson 1978a). Sterile brome grass (*Bromus sterilis*) is another annual of waste places in Britain with virtually no seed dormancy (Chancellor 1968). The lack of seed dormancy in *Vulpia* may explain why it is confined to the special environment of shingle and sand-dune habitats (Watkinson 1978b) and why it is not found in arable habitats, but *B. sterilis* has become a serious weed in arable fields where minimum tillage is practiced.

Despite the fact that the species represented most abundantly in seed pools often rely heavily or even entirely on this source of seeds for recruitment, the limited evidence available suggests that only a very tiny fraction of these seeds ever produce seedlings. Roberts and Ricketts (1979) found that the total weed seedling numbers present in fields after cultivation represented only 3–6 per cent of the weed seeds present in the top 10 cm of the soil before the weeds emerged. Naylor (1972) estimated the fraction of emerging seedlings of a weed grass *Alopecurus myosuroides* which were derived from seeds which had been in the soil longer than 1 year by a mark and recapture experiment. Seeds of the *A. myosuroides* were marked with a dilute solution of fluorescent paint and the seed production of plants in field plots was estimated. Quantities of marked seeds equal to one-tenth the number of naturally dispersed, unmarked seeds were added to the plots before the fields were harvested. When seedlings of *A. myosuroides* emerged in the experimental plots in the following season they were excavated to reveal the remains of the attached seed. The number of seedlings with a marked seed attached was equal to one-tenth of all the seedlings produced from seeds dispersed the previous year in the plot. Naylor calculated from this that 60–70 per cent of emerging seedlings were derived from the previous season's seed production. This figure suggests that, at least in this weed species, the seed pool does not provide many recruits to the growing population from seed that has been dormant for a long time.

There is an important difference between the cohort as defined in studies of animal populations and the way this term is used in practice for plant populations recruited from seed. Once an animal is born it generally embarks upon juvenile development straight away. Although there are some exceptions to this such as the extended diapause which may be found in insects, most metazoan populations possess no equivalent of the seed pool which may introduce a delay of indeterminate duration between the release of a seed and its emergence as a seedling. Thus, unlike an animal cohort, a cohort of emerging seedlings may consist of a collection of individuals which were not actually produced at the same time.

This may be important if there are several different genotypes among seeds and if natural selection is acting upon the population. In these circumstances different parental genotypes may contribute different quantities of seed to the seed pool in different years. When each new cohort of seedlings is drawn evenly from this pool of mixed origin it will contain genotypes derived from several different seasons and not just those most favoured in the most recent season. This effect of the seed pool may *buffer* genetic changes in plant populations (Templeton and Levin 1979). Epling, Lewis and Ball (1960) observed that this effect buffered year-to-year changes in the frequency of different flower colours in *Linanthus parryae*, an annual which occurs in the Mojave desert.

In other habitats where disturbance permits frequent recruitment from the seed pool, genetic variation may be maintained in the population because disturbance reduces competition from other plants and thus relaxes selection which tends to reduce genetic variability. Bosbach, Hurka and Haase (1982) found that populations of the annual weed shepherd's purse (*Capsella bursa-pastoris*) growing at disturbed sites contained more genetic variability than those experiencing less disturbance.

The fates of buried seeds

The fate of seeds in the soil is difficult to determine and comprehensive information on the dynamics of particular species' seed pools is sparse. In an experimental approach to the problem, Sarukhán (1974) sowed replicated samples of 100 viable seeds of three buttercups, *Ranunculus repens*, *R. bulbosus* and *R. acris*, into small areas of grassland in which natural seed dispersal was prevented. The fate of these seeds was classified in the five categories illustrated in Fig. 2.7 by counting emerging seedlings and retrieving buried seeds at intervals through the year. Seeds recovered from the soil were subjected to a germination test to detect dormancy and the ungerminable fraction was then tested for viability using tetrazolium chloride which stains living tissue red (Smith and Thorneberry 1951). The fates of seeds of *R. repens* and *R. bulbosus* determined by these methods over a 15-month period is shown in Fig. 2.8. Rodents were responsible for the heavy predation experi-

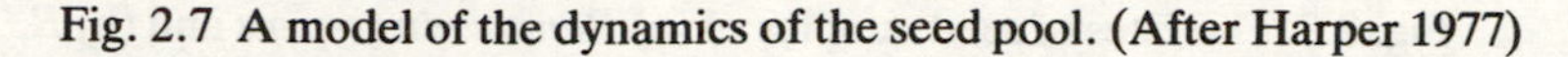

Fig. 2.7 A model of the dynamics of the seed pool. (After Harper 1977)

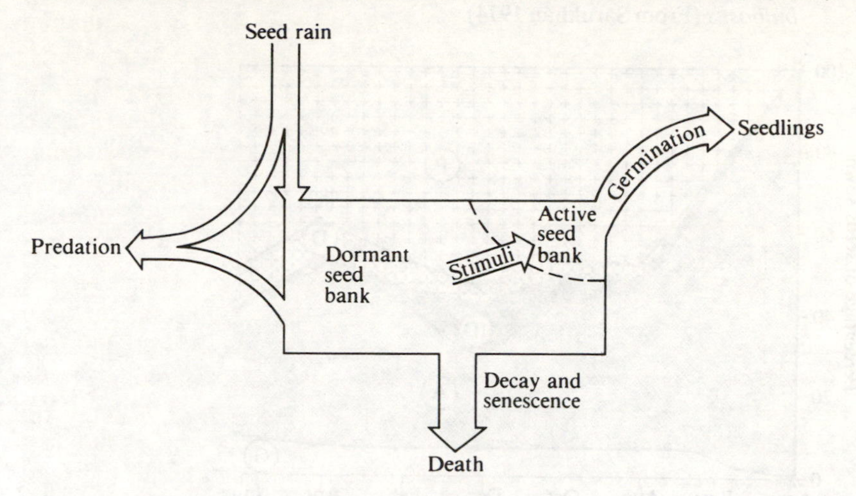

enced particularly by *R. repens* which lost 50 per cent of its seed pool in the first 6 months of the experiment. Other deaths occurred from unknown causes. Even though *R. repens* lost more seeds to predators than *R. bulbosus*, the seed pool of the latter species was depleted more rapidly as a result of substantial germination.

Seed predation by rodents is also an important factor in depleting the seed pool of the annual grasses which form some of the commonest species in Californian annual grassland. Borchert and Jain (1978) estimated the number of seeds taken by wild house mice (*Mus musculus*) and California voles (*Microtus californicus*) by excluding these predators from plots sown with known quantities of seeds of wild oats (*Avena fatua*), wild barley (*Hordeum leporinum*), ripgut brome (*Bromus diandrus*) and Italian ryegrass (*Lolium multiflorum*). *Avena fatua* seeds were preferred over those of the other species and 75 per cent of their seeds were eaten. Mice and voles consumed 44 per cent of the seeds of *H. leporinum* and 37 per cent of *B. diandrus*. The ultimate size of the population of growing plants in all of these species was not affected as severely as these levels of predation on seeds would suggest because a partly compensatory reduction in density-dependent mortality occurred in the plant populations from which seeds were removed by rodents. The survival from seeds to adult plants of *Avena, Hordeum* and *Bromus* was greater in the presence of seed predation than in its absence inside exclosure fences.

Granivorous rodents appear to be particularly important in depleting the seed pools of desert shrubs and herbs (Brown, Reichman and Davidson, 1979). An estimated 30–80 per cent of the seed losses observed by Nelson and Chew (1977) in a study in the Mojave desert were attributed to rodents, the most important species being the pocket

Fig. 2.8 The fates of seed samples of (a) *Ranunculus repens* and (b) *R. bulbosus* (From Sarukhán 1974)

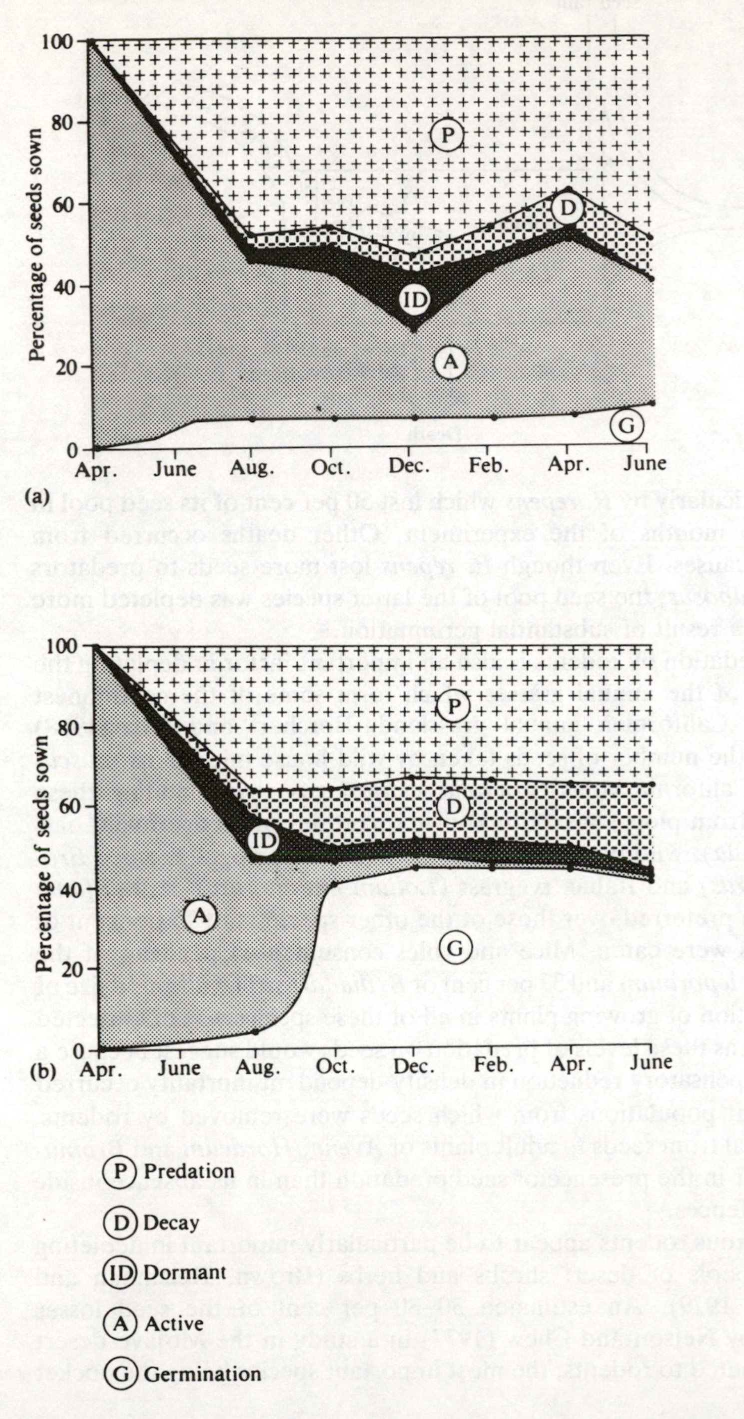

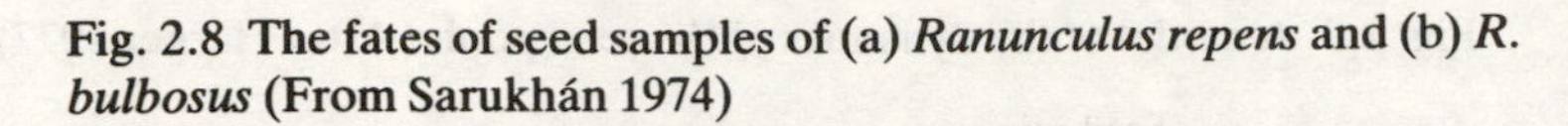

mouse (*Perognathus formosus*). Nelson and Chew estimated that only 25 per cent of seed losses were accounted for by germination, and they concluded that the total losses from the seed pool were not large enough to prevent a net accumulation of seeds in the soil. Harvester ants are also major predators of seeds in North American deserts (Brown, Reichman and Davidson 1979). Ants and rodents studied by Brown, Reichman and Davidson (1979) in the Chihuahua desert compete with each other for seeds. By selectively excluding rodents and ants from experimental plots, it was found that these seed predators both had marked, but different effects upon the community composition of annual plants, with ants taking smaller seeds than rodents (Inouye, Byers and Brown 1980). Granivores had both *direct* and *indirect* effects upon the abundance of annuals. Some species became more abundant following the removal of seed predators, but the increased density of these reduced the numbers of certain other species (Davidson, Samson and Inouye 1985).

Experiments on the decay of viable weed seed populations in cultivated soil by Roberts and Feast (1973) have shown that seed survivorship in the soil declines exponentially. However, seed survival is higher in uncultivated soil and these experiments may not reflect the situation for plants of other habitats. Odum (1978) selected 100 sites of abandoned settlements, abandoned farmland and demolished buildings in Denmark which were dominated by perennial vegetation and exposed the underlying soil to allow buried seeds to germinate. At virtually every site large numbers of annuals, biennials and other short-lived plants grew from the soil. The most spectacular example of seed survival in this study occurred in the soil removed from the excavation of an eleventh-century grave from which a plant of *Verbascum thapsiforme* germinated after 850 years of dormancy.

Seed dormancy

An enormous number of studies of seed dormancy have been conducted, mostly by physiologists attempting to elucidate the mechanisms which inhibit and trigger germination (Mayer and Polijakof-Mayber 1975). These studies have revealed a bewildering variety of factors influencing germination in different species including light intensity, photoperiod, light quality (spectral composition), temperature, temperature fluctuations, nitrates, O_2 and CO_2 levels, pH, moisture and physical abrasion of the seed coat, to name only the most common ones. To complicate matters still further, there is evidence that the conditions under which seeds are stored can induce dormancy in seeds which show no dormancy when freshly collected. There are also several examples of polymorphism within populations and geographic variation within species for seed dormancy. These show that many physiologically oriented studies carried out on seeds of unknown provenance, are of limited value to the population ecologist who would like to draw general conclusions from them. Most of the earlier studies of seed germination which are of interest to ecologists were carried out by weed researchers (Roberts 1970).

Despite these reservations about the evidence concerning seed dormancy in particular species, it is possible to sum up the general situation quite simply: 'Some seeds are born dormant, some acquire dormancy and some have dormancy thrust upon them' (Harper 1959). These three types of dormancy are termed *innate*, *induced* and *enforced* dormancy respectively (Harper 1977), and play slightly different roles in the regulation of germination. Fresh, innately dormant seed will not germinate in the conditions of normal germination tests (warmth and moisture on a filter paper or agar substrate) and will lie dormant in the seed pool till they receive a specific stimulus to break dormancy. Some period of 'after-ripening' may be required before the stimulus will have effect. Umbellifer seeds possess innate dormancy which is broken by a period of exposure to cold (stratification) when they are in the imbibed (water-saturated) state. Umbellifers such as cow parsley (*Anthriscus sylvestris*), hogweed (*Heracleum sphondylium*), wild angelica (*Angelica sylvestris*), pignut (*Conopodium majus*) and wild parsnip (*Pastinaca sativa*) which occur in Britain germinate in the spring after they have received the appropriate stratification and after the soil has warmed up sufficiently (Roberts 1979).

Dormancy may be induced in some seeds which are born without it by burial in the soil or by exposure to light filtered through a canopy of leaves. Wesson and Wareing (1969a,b) demonstrated that seeds of several species (*Chenopodium rubrum, Plantago lanceolata, Polygonum persicaria, Spergula arvensis* and others), which show no dormancy or germination requirement for light when freshly collected, occur in a dormant state in the seed pool and that this dormancy is broken by light when the soil is disturbed. A number of grassland species including several perennials produce seeds with little or no sign of innate dormancy when germinated in the dark but show induced dormancy when placed under a canopy of leaves (Silvertown 1980a) in response to the increased ratio of red/far-red light which occurs beneath leaves. These types of induced dormancy ensure that seeds do not germinate in circumstances where they will encounter a barrier of soil or vegetation which would reduce the survival of seedlings. In contrast to innate dormancy which locates seed germination in time, dormancy induced by burial or a leaf canopy locates or restricts seed germination in space as well as time.

Seeds in enforced dormancy may be released from this state simply by providing adequate water to allow imbibition at a normal temperature. Seeds held in enforced dormancy for a sufficient length of time may develop induced dormancy if they become buried, and combinations of innate, induced and enforced dormancy are also commonly found. Fresh seeds of knotgrass (*Polygonum aviculare*) show innate dormancy which prevents germination before the winter but this is broken by stratification and seedlings emerge in the spring. Seeds which have broken their innate dormancy but remain in enforced dormancy till the end of May then acquire renewed (induced) dormancy which prevents germination until another period of winter cold has passed (Courtney 1968; Roberts 1970). A similar seasonal adjustment of dormancy is achieved by a different means in fat hen (*Chenopodium album*) which produces non-dormant seeds early in the season and innately dormant ones in the main crop.

The buried seeds of many annual herbs go through yearly cycles of dormancy and non-dormancy induced by seasonal temperature changes (Baskin and Baskin 1985). Strict winter annuals such as mouse-ear cress (*Arabidopsis thaliana*) germinate best at low temperatures, summer annuals such as common ragweed (*Ambrosia artemisifolia*) germinate best at high temperatures. Cycles of induced dormancy occur in both these species but they are 6 months out of phase with each other (Fig. 2.9).

Many species show phenotypic polymorphism for seed dormancy and produce both dormant and non-dormant seeds in the same seed crop (Silvertown 1984). Arthur, Gale and Lawrence (1973) studied cohorts of seedlings produced from both dormant and non-dormant seeds of *Papaver dubium*, a poppy which is an arable weed in British fields. Seedlings emerging in the autumn from non-dormant seeds were prone to heavy mortality over the winter in some years, but the survivors of autumn cohorts produced larger plants and at least ten times more seeds than seedlings emerging in the spring.

Many annual species have a flush of seedlings in more than one season

Fig. 2.9 Annual cycles of induced dormancy in buried seeds of (a) the strict winter annual *Arabidopsis thaliana*; and (b) the summer annual *Ambrosia artemisifolia* retrieved from the soil and germinated at their respective temperature optima (see text). (From Baskin and Baskin 1983 and 1980 respectively)

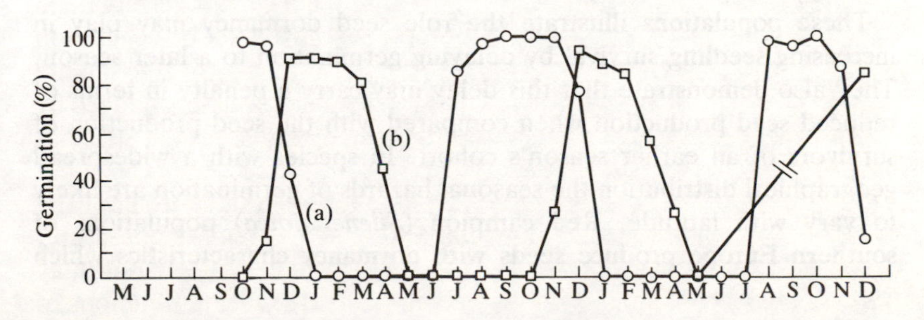

of the year. It may well be common in these populations for seed production to be relatively greater in the earliest cohort and seedling survival relatively greater in the later one. This pattern of seed production and seedling survivorship has been observed in autumn- and spring-germinating cohorts of prickly lettuce (*Lactuca serriola*) (Marks and Prince 1981) in Britain, and in summer and autumn cohorts of *Leavenworthia stylosa* in Tennessee cedar glades in the USA (Baskin and Baskin 1972). Autumn cohorts of *Lactuca* experience 1.3 times the mortality of spring ones because of winter temperatures. Summer cohorts of *Leavenworthia* experience over four times the mortality of autumn ones because of drought. Relative seed production is in favour of the earlier cohort in both species. Survivors of the autumn cohort of *Lactuca* produced twice the number of seeds produced by survivors of the spring cohort. Survivors of the summer cohort of *Leavenworthia* produce nearly eight times more seed per surviving plant than autumn cohorts.

The timing of germination is of crucial importance in the tropics too, though there seasonal rainfall rather than temperature is the important cue. In a study of timing of germination of approximately 185 dicot (mostly woody) species in the tropical forest of Barro Colorado Island (BCI) in Panama, Garwood (1982, 1983) found that over half the species delayed germination and that the length of the delay was roughly equal to the time between dispersal and the beginning of the rainy season. Forty per cent of species dispersed seeds in the rainy season and showed no dormancy. Another 18 per cent of plants dispersed seeds during the rainy season but delayed germination to the onset of the next one.

For pioneer trees, lianas, and canopy trees there was a single peak period of germination that occurred during the first 2 months of the rainy season, which lasts 8 months in all at BCI. Significantly, Garwood (1982) found that the dry-season mortality of these seedlings, which germinate in gaps, was related to time of emergence. Early-emerging seedlings survived better, apparently because these beat the competition and were then more able to tolerate drought. By contrast, germination of understorey and shade-tolerant species occurred throughout the rainy season, and in these there was no correlation between time of emergence and later dry-season mortality.

These populations illustrate the role seed dormancy may play in increasing seedling survival by delaying germination to a later season. They also demonstrate that this delay may carry a penalty in terms of reduced seed production when compared with the seed production of survivors of an earlier season's cohort. In species with a widespread geographical distribution the seasonal hazards of germination are likely to vary with latitude. Red campion (*Silene dioica*) populations of southern Europe produce seeds with dormancy characteristics which

Fig. 2.10 The percentage of seedlings surviving from cohorts of (a) *Androsace septentrionalis* emerging at 1-day intervals; and (b) *Tragopogon heterospermus* emerging at 10-day intervals in natural populations. (From Symonides 1977)

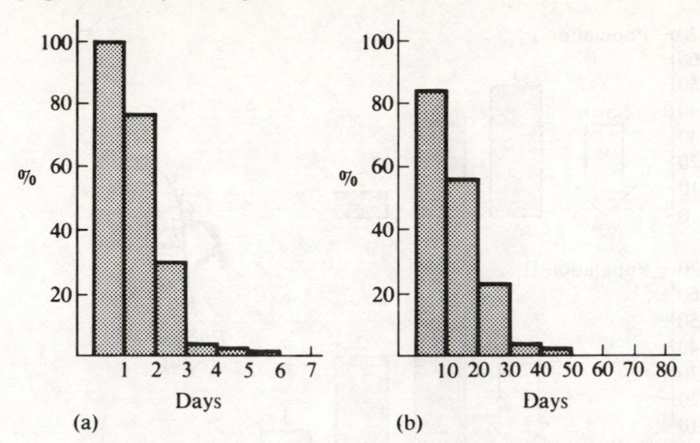

allow them to avoid summer drought, and seed populations from northern Europe are winter-dormant (Thompson 1975).

When the survivorship of seedlings emerging early and late within the same season's cohort is compared (Fig. 2.10), the first seedlings to emerge are generally at an advantage. The effects of even a slight delay in seedling emergence may be far-reaching for the subsequent fate of the plant. In a wood where he was studying the sweet white violet *Viola blanda*, Cook (1980) found a cohort of newly emerged seedlings in among a cohort he had marked 15 days before. He marked these also and then noted that the average size of seedlings from the late cohort remained consistently smaller than for the earlier ones during the following 3 years of their life. Smaller plants experienced a greater overall risk of mortality than larger ones. During a period of high mortality which occurred in the third year, plants of the later cohort suffered significantly greater mortality than plants of the earlier one which were less than 15 days their seniors in age.

Trends of this kind could either be due to environmental conditions deteriorating so that seedlings emerging later are adversely affected by the weather, or it may be due to older, larger seedlings capturing a larger share of resources and suppressing new ones as they emerge. Weaver and Cavers (1979) characterized these two alternatives as the effects of emergence *order* (e.g. whether seedlings are the first or last to emerge) and the effects of emergence *date* (e.g. March or April emergence) on seedling survival. They tested the relative importance of these two factors in populations of curl-leaved dock (*Rumex crispus*) and broad-leaved dock (*R. obtusifolius*) by sowing batches of seed at

Fig. 2.11 The percentage contribution of successive (1st, 2nd, 3rd) cohorts of seedlings of *Rumex crispus* to the final number of plants in three populations sown at different times. (Redrawn from Weaver and Cavers 1979)

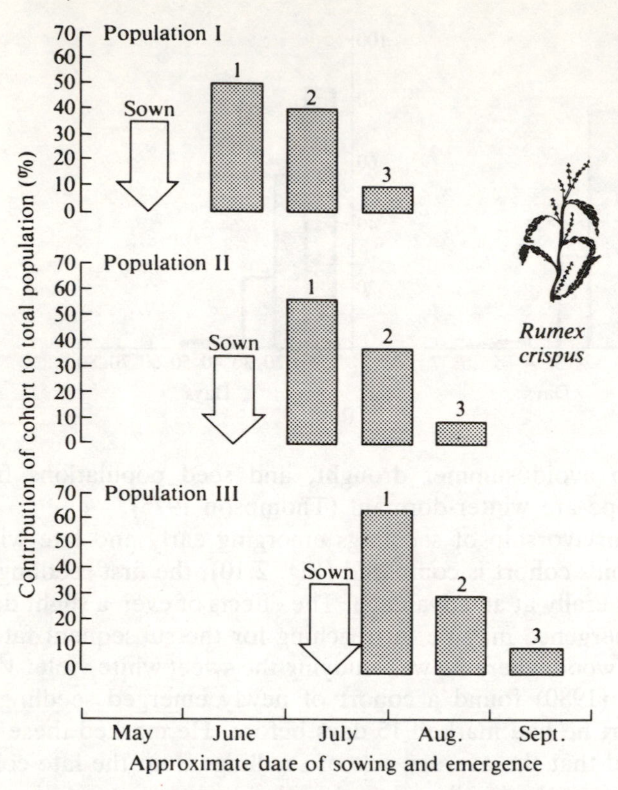

monthly intervals and comparing the relative contribution to the final dock population made by seedlings emerging in the first, second and third intervals in each period of seed germination. Percentage mortality was highest in the cohorts of later emergence order and their results (Fig. 2.11) showed that the order of emergence was more important than the date of emergence. This suggests that the accurate timing of seed germination and the rapid breaking of seed dormancy is essential to minimize the competition from other seedlings.

The safe site

A knowledge of the germination responses of seeds in the laboratory is not sufficient to predict accurately when and where a seed is capable of germination in the field. Simple ecological methods used in a number of studies have demonstrated that seed germination is highly responsive to fine-scale differences in the physical environment at the soil surface and that physiological studies of dormancy tell only half the story of why seeds do or do not germinate in the field. In a study of the response of seeds to microtopographical variation in the soil surface, Harper, Williams and Sagar (1965) sowed a seed bed with equal proportions of the seeds of three plantains *Plantago media*, *P. lanceolata*, and *P. major*. The seed bed was divided into sub-plots and various objects were then placed on the surface of these. The treatments, listed in the legend of Fig. 2.12, had selective effects on the seedling emergence of the different species which showed marked differences in their response to different conditions of soil microtopography and micro-environment.

Harper, Williams and Sagar (1965) adopted the term *safe site* to describe those specific conditions in the soil surface which permit seeds to escape all the hazards of the pre-germination phase (including predation) and to overcome dormancy. Although this term shares with the term *niche* the unfortunate property that a particular species' safe site can only be identified *after* it has been successfully occupied, the term provides a useful conceptual handle with which to grapple with the idea that a multitude of factors affect seed germination.

The shape and size of a seed in relation to the soil particles of the surface in which it rests seem to play an important part in determining the availability of the water needed for successful germination. Oomes and Elberse (1976) compared the seed germination of six grassland species when sown on an even soil surface and in 10 mm and 20 mm wide grooves in the surface of the soil. Yarrow (*Achillea millefolium*) has flat seeds which germinated best on the even surface, but several species with other shapes (e.g. ox-eye daisy *Chrysanthemum leucanthemum* and self-heal *Prunella vulgaris*) germinated poorly when lying in this position and did much better in the 20 mm grooves.

Perhaps one of the most important factors which determines the spatial pattern of seedling emergence and recruitment in the field is the microdistribution of the leaves and leaf canopies of established plants. Oak seedlings of *Quercus robur* are light-demanding and are also prone to defoliation by moth caterpillars which can descend on them from established trees. Either or both of these factors may explain why oak saplings are not often found beneath the canopy of large oaks (Fig. 2.13). Interactions with a similar result also occur among herbaceous plants (Silvertown 1981a).

Decaying logs may act as safe sites, or 'nurse logs' for tree seedlings. In Hawaii the native name for the tree fern is 'Mother of Ohia', because Ohia (*Metrosideros collina*) germinates on its fallen trunks and even on living trees. In one 200 m square plot in the Cascade mountains in Oregon Christy and Mack (1984) found that 98 per cent of juveniles of western hemlock *Tsuga heterophylla* occurred on decaying logs of *Pseudotsuga menziesii*, even though these covered only 6 per cent of the area. In experiments they found that seeds germinated on many types of log and

Fig. 2.12 (a) The distribution of various treatments to the *Plantago* seed bed: (1) and (2) two kinds of depression in the soil surface; (3) a sheet of glass laid on the soil surface; (4a) and (4b) sheets of glass placed vertically in the soil; (5), (6) and (7) rectangular wooden frames of three different depths pressed into the soil surface; x worm casts. (b) the distribution of seedlings of *P. lanceolata*, (c) *P. media* and (d) *P. major*. (From Harper, Williams and Sagar 1965)

(*b*) *Plantago lanceolata*

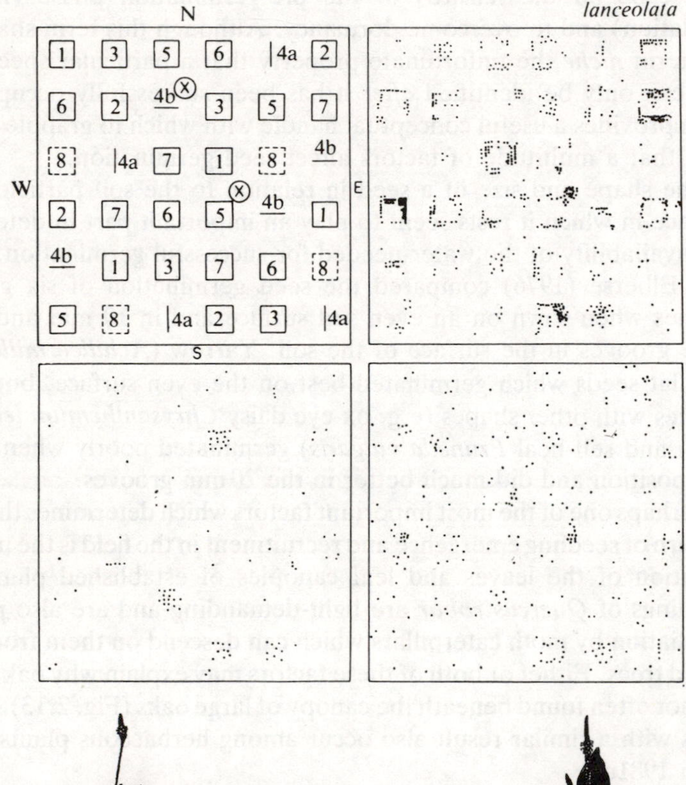

(*c*) *Plantago media*

(*d*) *Plantago major*

Fig. 2.13 The distribution of young oaks around a mature tree at Silwood Park in Berkshire. (From Mellanby 1968)

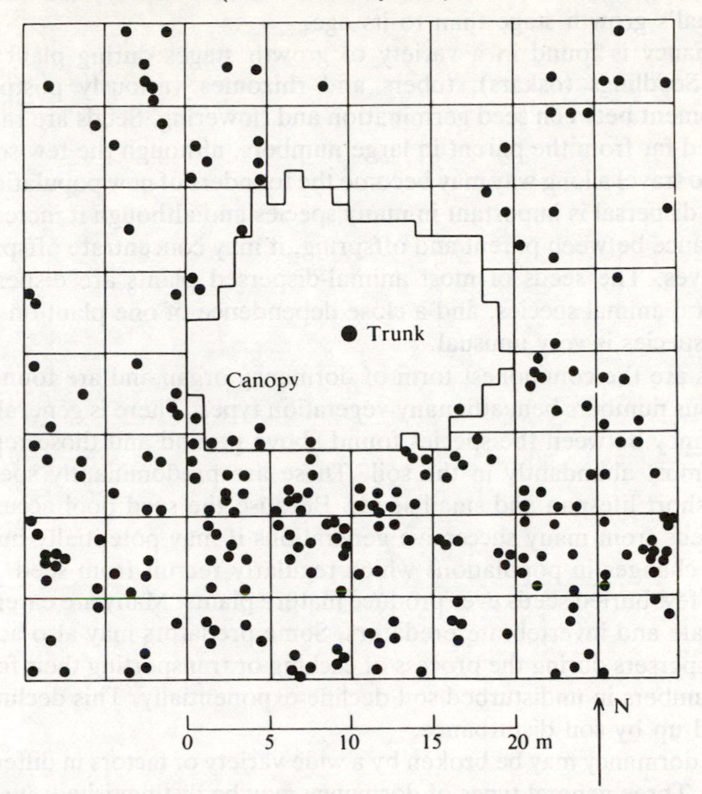

on mineral soil; however seedling survival was better on *Pseudotsuga* logs with rotten heartwood than on other substrata. Logs raise seedling *Tsuga* out of the litter which tends to bury and kill them, but the logs must be sufficiently decayed to allow seeds to lodge and seedlings to establish, yet not so rotten as to allow litter to accumulate on them. *Picea sitchensis* and *Thuja* also establish on nurse logs in the Pacific Northwest.

Now that we have dealt with some of the biological aspects of plant populations and their description in life tables, we are ready to apply some simple population models to them in the next chapter.

Summary

A *life table* lists the age-specific probabilities of mortality and survival for a cohort of individuals. A *fecundity schedule* lists age-specific reproductive rates in parallel to the life table. *Dynamic* life tables are drawn up by following the fate of a cohort. *Static* life tables are derived by more indirect methods.

Because of the plastic growth of plants, individuals of the same *chronological age* may be at quite different *stages* of growth. Particularly in monocarpic plants, flowering is often more closely tied to an individual's growth stage than to its age.

Dormancy is found in a variety of growth stages during plant life cycles. Seedlings (oskars), tubers and rhizomes variously postpone development between seed germination and flowering. Seeds are rarely dispersed far from the parent in large numbers, although the few seeds which do travel a long way may become the founders of new populations. Animal dispersal is important in many species and although it increases the distance between parent and offspring, it may concentrate offspring themselves. The seeds of most animal-dispersed plants are dispersed by several animal species, and a close dependence of one plant on one animal species is very unusual.

Seeds are the commonest form of dormancy organ and are found in enormous numbers beneath many vegetation types. There is generally a discrepancy between the species found above ground and those represented most abundantly in the soil. These are predominantly species with a short lifespan and small seeds. Because the seed pool accumulates seeds from many successive generations it may potentially buffer genetic changes in populations which regularly recruit from seed.

Very few buried seeds ever produce mature plants. Many are eaten by vertebrate and invertebrate predators. Some predators may also act as seed dispersers during the process of caching or transporting their food. Seed numbers in undisturbed soil decline exponentially. This decline is speeded up by soil disturbance.

Seed dormancy may be broken by a wide variety of factors in different species. Three general types of dormancy may be distinguished: *innate*, *induced* and *enforced*. Some of these may occur alone, successively in time or in the same seed and in different seeds in the same population. The breaking of dormancy and the timing of germination have measurable effects upon the subsequent survival and reproduction of seedlings.

Various physical environmental factors determine whether a seed will germinate in the soil and how the surviving seedlings will be distributed. The term *safe site* has been given to those conditions which permit successful establishment of a particular species from seed.

3
Simple population models

Models of how a population behaves are needed to understand how the fundamental demographic processes of birth, death, immigration and emigration contribute to changes (or to stability) in population size. We have already met the simplest population model in Chapter 1:

$$N_{t+1} = N_t + B - D + I - E \quad [3.1]$$

It is worth returning to this equation, which must be the nearest thing there is to an algebraic truism, whenever the going gets tough with more complex descriptions of population change. However, the mathematics in this book is kept to a minimum and should give the reader little cause for worry. More complex equations for population change arise from attempts to account for the way in which birth-*rates*, death-*rates* and migration-*rates* alter with population density and age structure and with the effects of competitors, predators, pathogens and mutualists. A thorough mathematical treatment of these relationships is beyond the scope of this book, but we will deal with the simplest models of the effects of density and age structure in this chapter.

Difference equation models

Equation [3.1] is technically known as a *difference equation*. Models based upon this equation are appropriate for populations in which generations are discrete and do not overlap. Some annual plants fall into this category. The net rate at which a population is increasing or decreasing is called the *net reproductive rate* R_0, and is simply the ratio of offspring produced in generation N_{t+1} to the population size in the previous generation N_t:

$$R_0 = N_{t+1} / N_t \quad [3.2]$$

The net reproductive rate (R_0) is the *slope* of the line on a graph of N_{t+1} plotted against N_t. A population is at equilibrium when there is no net change in numbers, though there are births, deaths etc. Then, $N_{t+1} = N_t$, and $R_0 = 1$. The equilibrium line $R_0 = 1$ is represented by the diagonal in Fig. 3.1(a). If $R_0 > 1$, the population will increase geometrically by a factor of R_0 in each time interval. This situation has been observed on some occasions (Fig. 3.2(a and b)), but obviously cannot occur anywhere

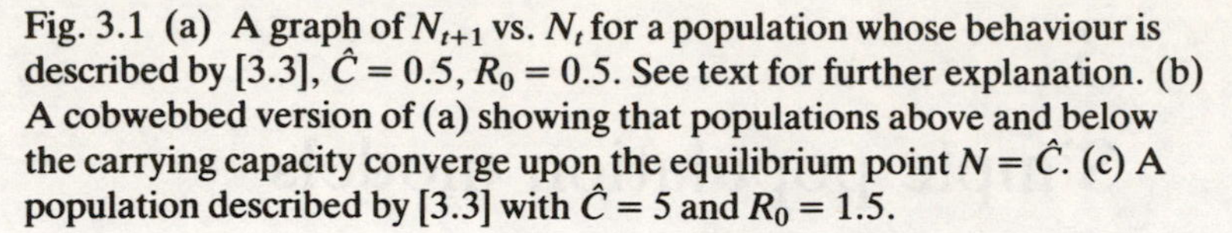

Fig. 3.1 (a) A graph of N_{t+1} vs. N_t for a population whose behaviour is described by [3.3], $\hat{C} = 0.5$, $R_0 = 0.5$. See text for further explanation. (b) A cobwebbed version of (a) showing that populations above and below the carrying capacity converge upon the equilibrium point $N = \hat{C}$. (c) A population described by [3.3] with $\hat{C} = 5$ and $R_0 = 1.5$.

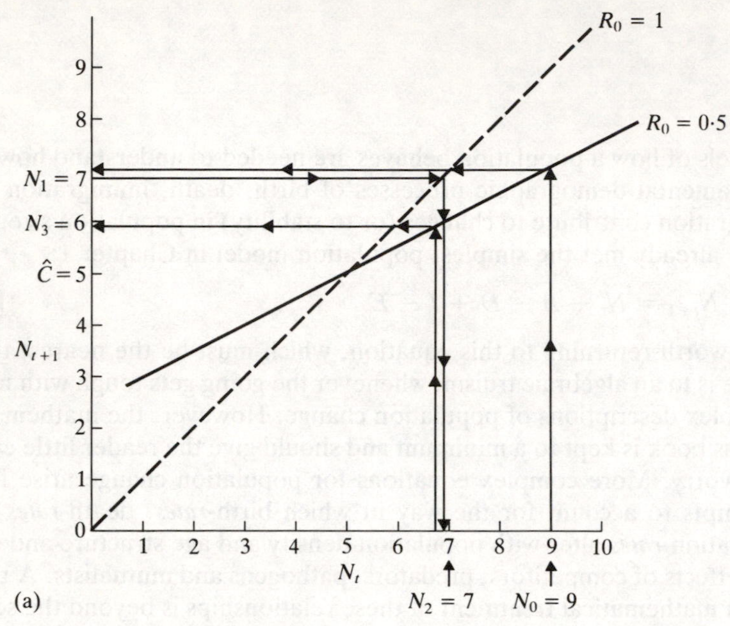

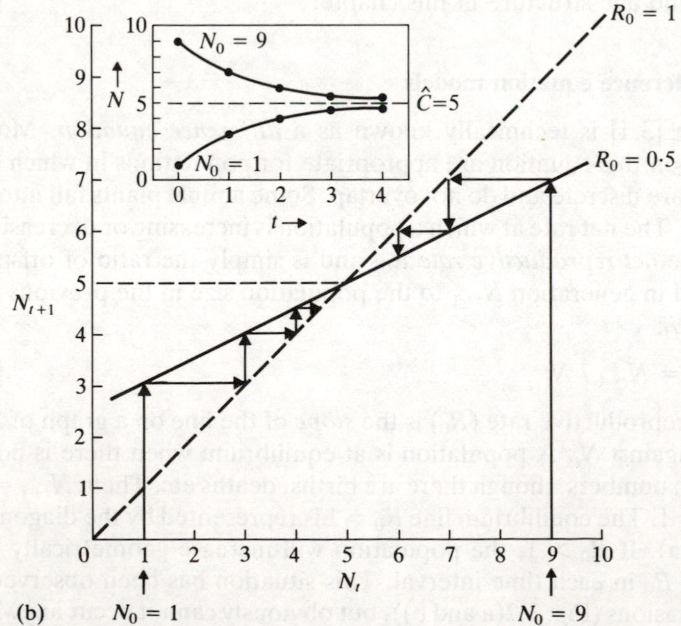

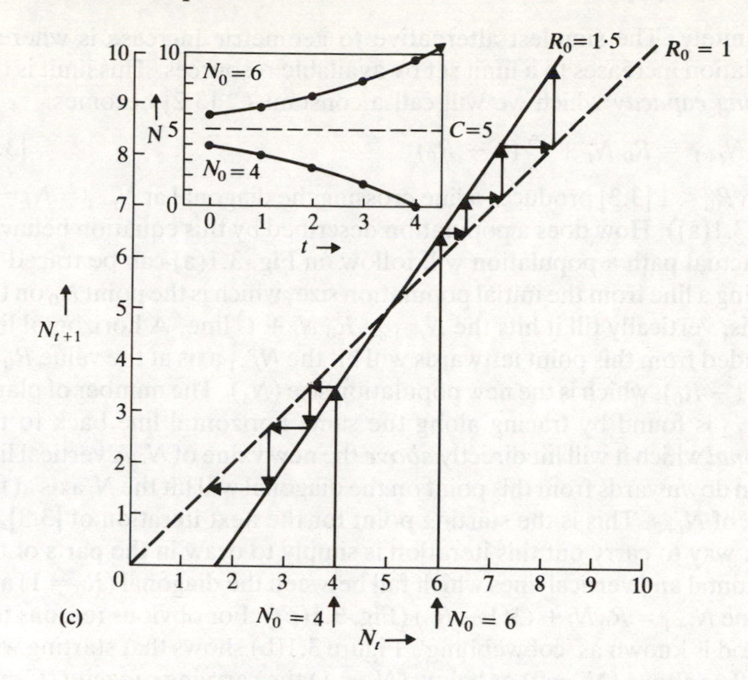

Fig. 3.2 The exponential increase of (a) *Avena fatua* infesting a barley crop at Boxworth Experimental Farm, Cambridge, England; and (b) of *Pinus sylvestris* between 9500 and 9000 years ago when this tree was invading Hockham Mere, Norfolk, England. The abundance of *P. sylvestris* 9500–9000 years ago is plotted as the density of pollen grains in peat samples. The density of pollen should be correlated with the historical abundance of the tree. ((a) data from Selman 1970; (b) from Bennett 1983)

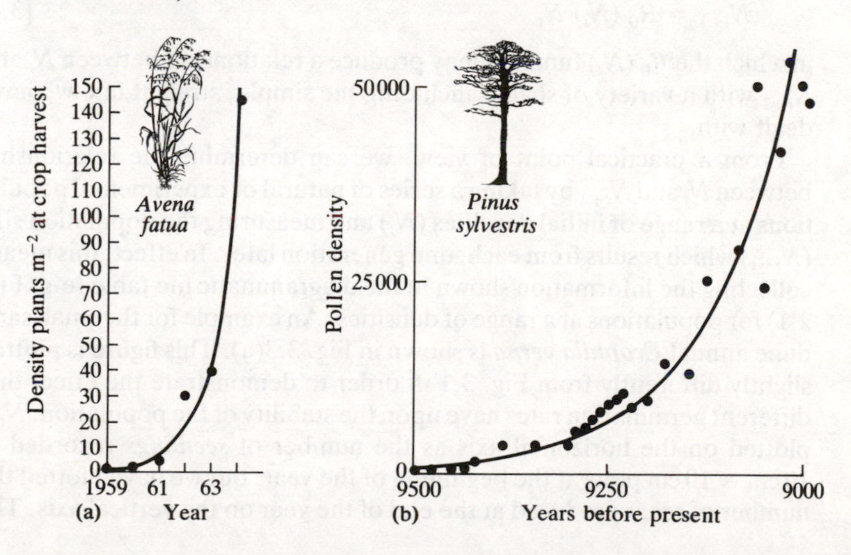

indefinitely. The simplest alternative to geometric increase is where a population increases to a limit set by available resources. This limit is the *carrying capacity* which we will call a constant $\hat{C}$. [3.2] becomes:

$$N_{t+1} = R_0 N_t + \hat{C} (1 - R_0) \quad [3.3]$$

When $R_0 < 1$ [3.3] produces a line crossing the diagonal at $N_{t+1} = N_t = \hat{C}$ (Fig. 3.1(a)). How does a population described by this equation behave? The actual path a population will follow on Fig. 3.1(a) can be traced by drawing a line from the initial population size, which is the point N_0 on the N_t axis, vertically till it hits the $N_{t+1} = R_0 N_t + \hat{C}$ line. A horizontal line extended from this point leftwards will hit the N_{t+1} axis at the value $R_0 N_t + \hat{C}(1 - R_0)$, which is the new population size (N_1). The number of plants at N_{t+2} is found by tracing along the same horizontal line back to the *diagonal* which it will hit directly *above* the new value of N. A vertical line drawn downwards from this point on the diagonal will hit the N_t axis at the value of N_{t+2}. This is the starting point for the next iteration of [3.3]. A quick way to carry out this iteration is simply to draw in the parts of the horizontal and vertical lines which fall between the diagonal ($R_0 = 1$) and the line $N_{t+1} = R_0 N_t + \hat{C}(1 - R_0)$ (Fig. 3.1(b)). For obvious reasons this method is known as 'cobwebbing'. Figure 3.1(b) shows that starting with N_0 either above ($N_0 = 9$) or below ($N_0 = 1$) the carrying capacity ($\hat{C} = 5$) causes the population to converge on the intersection of the line for [3.3] and the diagonal. This will occur whenever $R_0 < 1$. Such populations have a stable equilibrium at a density of C individuals (Fig. 3.1(b) inset).

When $R_0 > 1$ the population has an unstable equilibrium at $\hat{C}$ (Fig. 3.1(c)). In reality, R_0 is unlikely to have a constant value and will change with the density of the population. The general equation in which R_0 is a function of N_t is:

$$N_{t+1} = R_0 (N_t) N_t \quad [3.4]$$

in which the $R_0 (N_t)$ function may produce a relationship between N_t and N_{t+1} with a variety of shapes including the simple, straight line we have dealt with.

From a practical point of view, we can determine the relationship between N_t and N_{t+1} by taking a series of natural or experimental populations at a range of initial densities (N_t) and measuring the population size (N_{t+1}) which results from each, one generation later. In effect, this means collecting the information shown in the diagrammatic life table (e.g. Fig. 2.1) for populations at a range of densities. An example for the small sand dune annual *Erophila verna* is shown in Fig. 3.3(a). This figure is plotted slightly differently from Fig. 3.1 in order to demonstrate the effect that different germination rates have upon the stability of the population. N_t is plotted on the horizontal axis as the number of *seedlings* recorded in 10 cm × 10 cm plots at the beginning of the year, but we have plotted the number of *seeds* produced at the end of the year on the vertical axis. The

Fig. 3.3 The relationship of seed output to initial seedling density in a dune population of *Erophila verna* in Poland and the dynamics of the population predicted by cobwebbing when the germination success of seeds is (a) 0.5%; (b) 1%; and (c) 2%. See text for further details. (Data from Symonides 1983a, b)

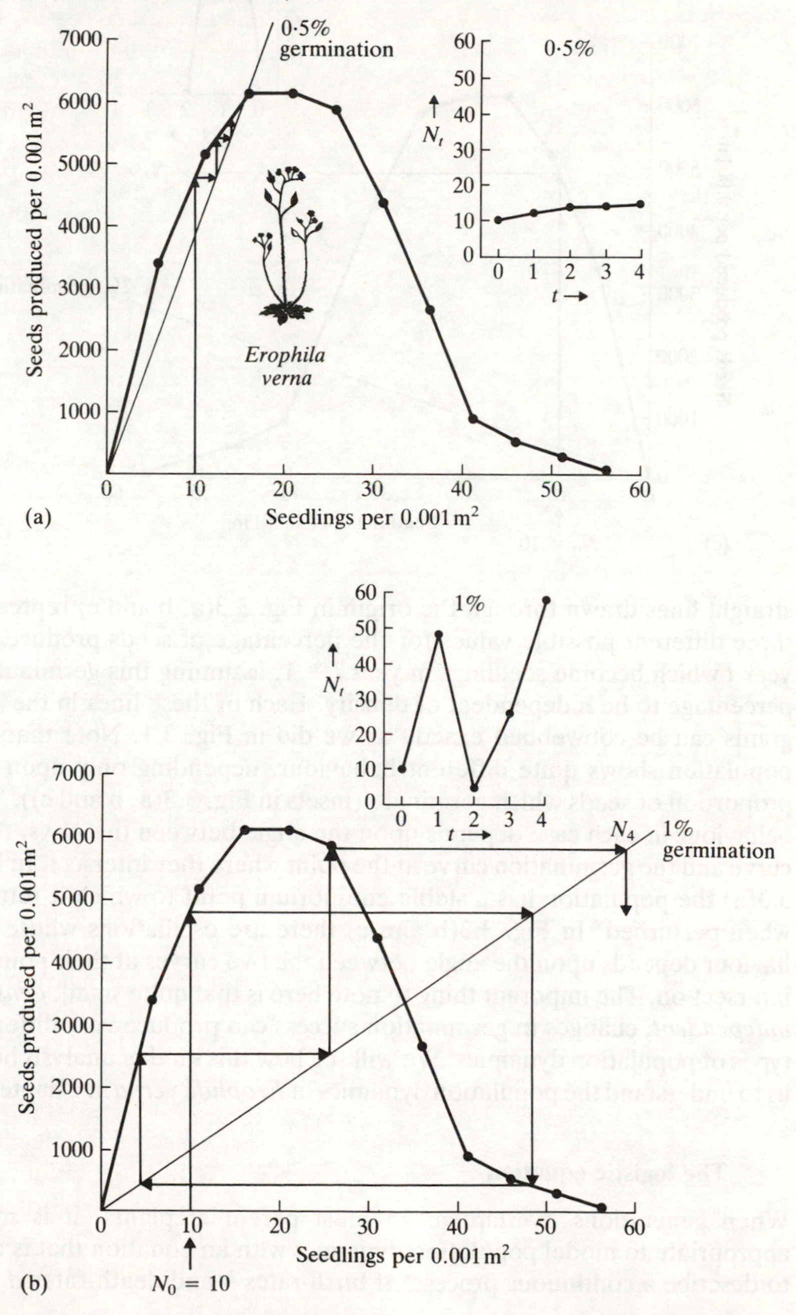

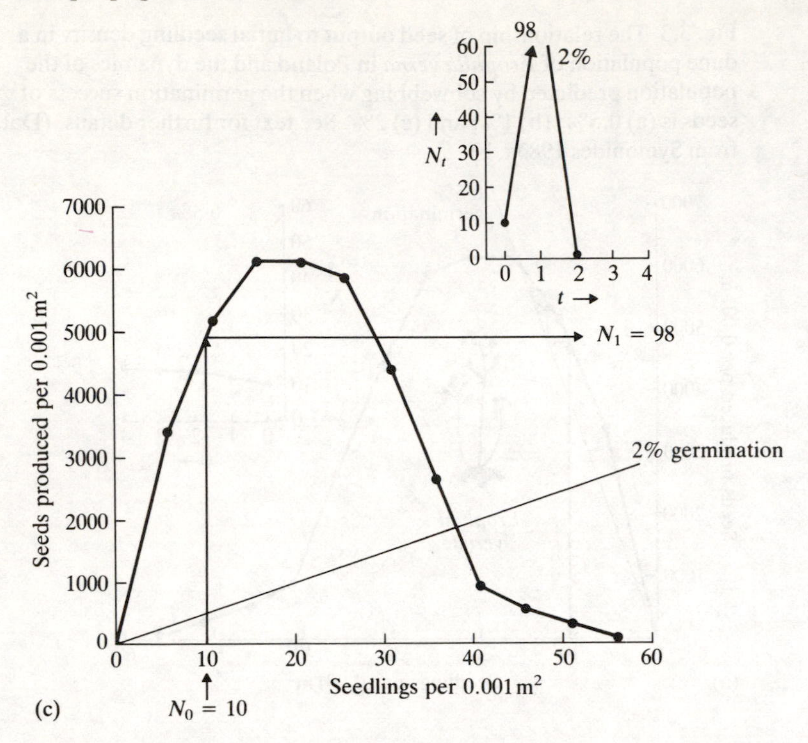

straight lines drawn through the origin in Fig. 3.3(a, b and c) represent three different possible values for the percentage of seeds produced in year t which become seedlings in year $t + 1$, assuming this germination percentage to be independent of density. Each of these lines in the diagrams can be cobwebbed exactly as we did in Fig. 3.1. Note that the population shows quite different behaviour, depending only upon the proportion of seeds which germinate (insets in Fig. 3.3(a, b and c)). The behaviour in each case depends upon the angle between the N_t vs. N_{t+1} curve and the germination curve at the point where they intersect. In Fig. 3.3(a) the population has a stable equilibrium point to which it returns when perturbed. In Fig. 3.3(b and c) there are oscillations whose behaviour depends upon the angle between the two curves at their point of intersection. The imporant thing to note here is that quite small, *density-independent*, changes in germination success can produce very different types of population dynamics. We will see how this kind of analysis helps us to understand the population dynamics of *Erophila verna* in Chapter 4.

The logistic equation

When generations overlap, as in most perennial plants, it is more appropriate to model population dynamics with an equation that is able to describe a continuous process. If birth-rates b and death-rates d in a population are constant we may calculate the *instantaneous rate* of increase r from these two parameters;

$$r = b - d \tag{3.5}$$

The rate of change of population size dN/dt is then given by the differential equation:

$$\frac{dN}{dt} = rN \tag{3.6}$$

where N is the size of the population at a particular instant in time. To obtain the size of the population N_t at some time t, [1.4] is integrated to give:

$$N_t = N_0 e^{rt} \tag{3.7}$$

where N_0 is the initial size of the population at some time designated zero, e is a constant = 2.718 (the base of natural logarithms), r is the intrinsic rate of increase as defined in [3.5] and t is the time elapsed since time zero. The population growth described by [3.7] is *exponential*, which means that population size goes on doubling (or halving if $d > b$) at a constant rate as shown in Fig. 3.2. Equation [3.6] can be modified to take account of the limitations to population growth which operate as population size increases towards the limits of resource availability, if we

Fig. 3.4 Increase of a moss population colonizing bare rock on the Icelandic island of Surtsey. (From Fridrickson 1975)

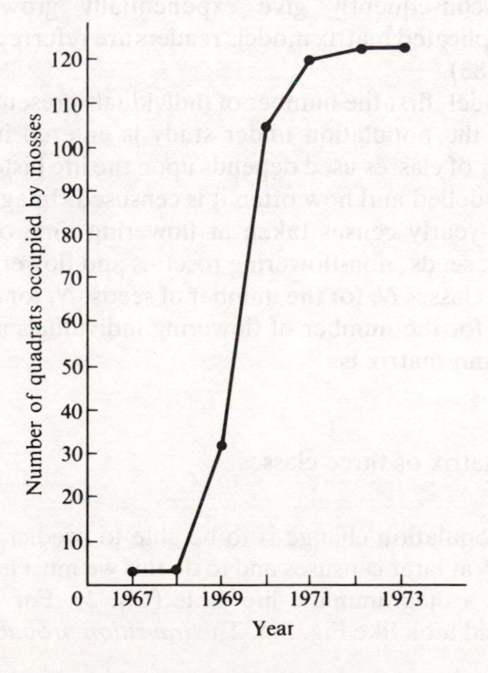

add a term which reduces rN as that limit, the carrying capacity K, is reached:

$$\frac{dN}{dt} = rN\frac{(K - N)}{K} \qquad [3.8]$$

This is the *logistic* equation. It approximately describes the colonization of new rock surface by mosses on the volcanic island of Surtsey (Fig. 3.4), perhaps one of the simplest situations an ecologist could ever hope to study.

The logistic equation is based upon several assumptions: 1. that population increase is *instantaneous*, 2. that all individuals are exactly equivalent and 3. that there is no spatial structure to the population. The logistic equation is a useful basis for certain population models because of its simplicity, but we will see in Chapter 9 how a model of interspecific competition based upon the logistic comes unstuck when the assumptions of the equation are violated.

Matrix models

Matrix models allow us to describe the behaviour of populations which have overlapping generations, and which have individuals that fall into different age or size classes and have rates of reproduction and death which depend upon age and/or size. These models are more realistic than models based upon the logistic equation and they are simple to use, especially with a microcomputer. The models used here do not involve density-dependence and consequently give exponentially growing populations. For more complicated matrix models readers are referred to Pielou (1977) and Law (1983).

To construct a matrix model, first the number of individuals present in each stage or age class in the population under study is entered in a *column matrix*. The number of classes used depends upon the life history of the population being modelled and how often it is censused. Imagine the simple case of a once-yearly census taken at flowering time of a perennial with three stages: seeds, non-flowering rosettes and flowering individuals. If we call these classes N_s for the number of seeds, N_r for the number of rosettes and N_f for the number of flowering individuals at a particular census, the column matrix is:

$$\begin{bmatrix} N_s \\ N_r \\ N_f \end{bmatrix}$$

Matrix 1.
A column matrix of three classes

The object of modelling population change is to be able to predict the magnitude of N_s, N_r, and N_f at later censuses and to do this we must have the information to fill out a diagrammatic life table (Ch. 2). For our hypothetical plant this would look like Fig. 3.5. The *transition probabili-*

Fig. 3.5 Diagrammatic life table for the hypothetical perennial plant discussed in the text.

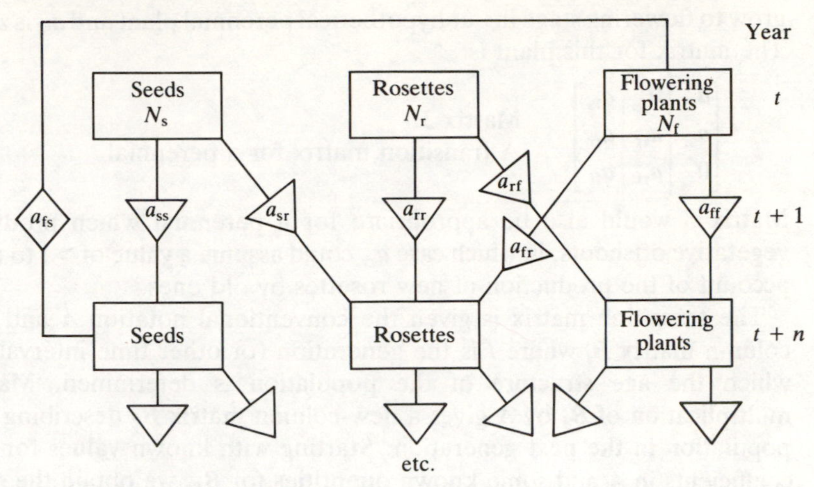

ties in Fig. 3.5 which quantify the fate of each stage in the plants' life cycle are entered in a *transition matrix*. This matrix is square, with each side the same length as the column matrix, so that the transition matrix can accommodate coefficients for the probability of every transition from one age class to another:

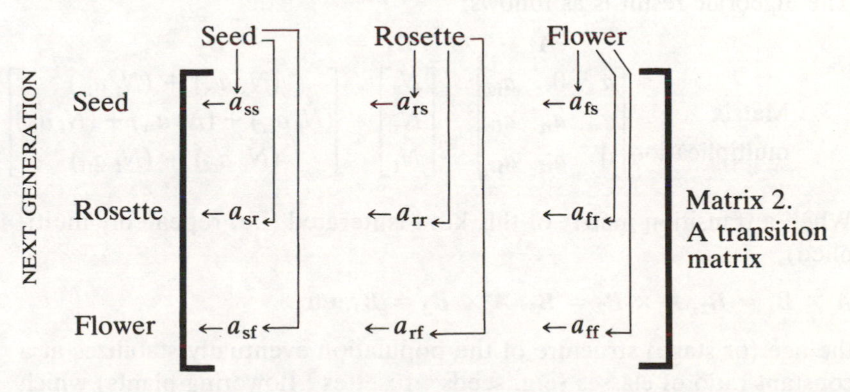

In matrix 2, a_{ss} is the probability that a seed this year will not germinate but will remain a viable seed next year, a_{sr} is the probability that the seed will become a rosette next year and a_{sf} is its probability of flowering and so on. The first column gives the fate of seeds, the second the fate of rosettes and the third the fate of flowering plants. Some of the coefficients in the transition matrix have no biological meaning, for instance by definition non-flowering rosettes cannot produce seeds and a_{rs} is replaced by a zero when we write out the values of the matrix. Other coefficients in the matrix may be zero because of the peculiarities of the life history of the population. For example, seeds take more than 1 year to grow to flowering stage in our hypothetical perennial plant and a_{sf} is zero. The matrix for this plant is:

$$\begin{bmatrix} a_{ss} & 0 & a_{fs} \\ a_{sr} & a_{rr} & a_{fr} \\ 0 & a_{rf} & a_{ff} \end{bmatrix}$$

Matrix 3.
A transition matrix for a perennial

Matrix 3 would also be appropriate for a perennial which produces vegetative offshoots, in which case a_{rr} could assume a value of >1 to take account of the production of new rosettes by old ones.

The transition matrix is given the conventional notation $\boldsymbol{A}$ and the column matrix $\boldsymbol{B}_t$ where t is the generation (or other time interval) at which the age structure of the population is determined. Matrix multiplication of $\boldsymbol{B}_1$ by $\boldsymbol{A}$ gives a new column matrix $\boldsymbol{B}_2$ describing the population in the next generation. Starting with known values for the coefficients in $\boldsymbol{A}$ and some known quantities for $\boldsymbol{B}_1$, we obtain the new value of N_s in $\boldsymbol{B}_2$ by multiplying each coefficient (a_{ss}, 0, a_{fs}) in the first row of $\boldsymbol{A}$ by the corresponding value (N_s, N_r, N_f) in $\boldsymbol{B}_1$ and then summing the products. The new value of N_r is obtained by multiplying each coefficient (a_{sr}, a_{rr}, a_{fr}) in the second row of $\boldsymbol{A}$ by the corresponding value (N_s, N_r, N_f) in $\boldsymbol{B}_1$ and summing the products and so on. It may be useful at this point to refer back to the explanatory diagram of a transition matrix (matrix 2) for the biological meaning of the multiplication $\boldsymbol{A} \times \boldsymbol{B}_1$. The algebraic result is as follows:

Matrix multiplication

$$\underset{\boldsymbol{A}}{\begin{bmatrix} a_{ss} & 0 & a_{fs} \\ a_{sr} & a_{rr} & a_{fr} \\ 0 & a_{rf} & a_{ff} \end{bmatrix}} \times \underset{\boldsymbol{B}_1}{\begin{bmatrix} N_s \\ N_r \\ N_f \end{bmatrix}} = \underset{\boldsymbol{B}_2}{\begin{bmatrix} (N_s\, a_{ss}) + (N_f\, a_{fs}) \\ (N_s\, a_{sr}) + (N_r\, a_{rr}) + (N_f\, a_{fr}) \\ (N_r\, a_{rf}) + (N_f\, a_{ff}) \end{bmatrix}}$$

When a transition matrix of this kind is iterated (i.e. repeatedly multiplied);

$$\boldsymbol{A} \times \boldsymbol{B}_1 = \boldsymbol{B}_2,\ \boldsymbol{A} \times \boldsymbol{B}_2 = \boldsymbol{B}_3,\ \boldsymbol{A} \times \boldsymbol{B}_3 = \boldsymbol{B}_4, \text{ etc.}$$

the age (or stage) structure of the population eventually stabilizes at a constant ratio of classes (e.g. seeds : rosettes : flowering plants) which depends upon the values of the coefficients in $\boldsymbol{A}$ and is independent of the values in $\boldsymbol{B}_1$. Once $\boldsymbol{A} \times \boldsymbol{B}_1$ has been iterated to the point at which a stable age distribution is reached, the ratio of any one age class (say N_s) to the same age class in the next generation gives the per-year rate of increase $\lambda = R_0/t$, where t is the length of a generation. For example for an annual plant $t = 1$ and $\lambda = R_0$.

The deliberate emphasis in this section has been on matrices as a tool rather than on the biological meaning and experimental derivation of transition probabilities and age structures. The great advantage of matrix models is that they allow us to determine the effect of changing transition probabilities such as those between the stages seed → rosette or between flowering plant → seed.

Sensitivity analysis

The main feature of a matrix model is that it explicitly takes into account the contribution made by individuals in different age or stage classes to the dynamics of a population. This is important because individuals of different age/stage contribute differently to the yearly rate of increase (lambda). Sensitivity analysis is a technique for calculating the relative effect that making small changes in the value of the different transition probabilities (the elements in the **A** matrix) would have upon lambda. Two things are required: 1. the stable age distribution calculated by iteration of the transition matrix; and 2. the *reproductive value* V_x of individuals in each age/stage class. The reproductive value of an individual is simply the relative contribution a typical individual of its class will make to the next generation, taking into account its life expectancy and future fecundity. (We will come across reproductive value again in Chapter 7 but note that the formula given there for calculating it is valid only when lambda = 1, and so is inappropriate here.)

The sensitivity of lambda to small changes in the transition probability between classes *i* and *j* (in matrix notation a_{ij}) is calculated by multiplying the reproductive value of an individual in class *i* by the percentage of all individuals found in class *j* of the stable age structure (Caswell 1978, 1986). This technique was used by Mortimer (1983) to determine the relative importance of transitions between life-cycle stages in the perennial grass weed *Elymus repens*. This plant multiplies by both vegetative buds and seeds and was analysed using a matrix model with four classes: seeds, adults <1 year old, adults 1–2 yr, adults >2 yr and buds. The per year rate of increase (lambda) of this population was 2.96. The sensitivity matrix (Table 3.1) shows that this rate would have been equally sensitive

Table 3.1 Sensitivity matrix for a population of *Elymus repens* showing the relative sensitivity of the annual rate of population increase to changes in the most important transitions between stages in the life cycle. (From Mortimer 1983)

	Seed	Adults <1 yr	Adults 1–2 yr	Adults >2 yr	Buds
Seed					
Adults <1 yr	3.71				
Adults 1–2 yr		0.65			
Adults >2 yr			0.18		
Buds		3.69	1.07	1.89	

to changes in seedling establishment (3.71) as to changes in bud production by young (<1 yr) adults (3.69) but, in total, bud production by all adults (3.69 + 1.07 + 1.89) was more important than seedling establishment.

Some caution is needed when interpreting sensitivity values because they are calculated on the assumption that small perturbations to all transition probabilities are equally likely. This ignores the fact that, for example, seed production may tend to vary more than adult survival in a perennial plant. Even if lambda is equally sensitive to small perturbations in both these life cycle stages, the fact that perturbations are larger or occur more often at one stage than the other will mean, *in practice*, that seed production is the more important variable in determining lambda. Sensitivity values need to be calibrated to allow for this effect (van Groenendael 1985). We will see several uses of matrix models and sensitivity analysis in the following chapters.

Population models demonstrate that relationships between density and mortality and between density and fecundity are important in the dynamics of plant populations. These relationships are dealt with in the next chapter.

Summary

Models of population dynamics are needed to understand how birth, death, immigration and emigration contribute to changes in population size. The simplest model [3.1] is a *difference equation* which is appropriate for populations with *discrete (non-overlapping) generations*. When density dependence is incorporated in the form of a *carrying capacity* these models can be quite realistic for annual plants. Density-independent factors may affect the behaviour and *stability* of such populations.

Models based upon the *logistic equation* [3.8] are appropriate for populations in which generations overlap. Such models are based upon the assumption that, when density dependence is not operating, population increase occurs at an *instantaneous rate r*.

Matrix models explicitly incorporate the age- or stage-structure of a population with overlapping generations, but cannot easily deal with density dependence. The *sensitivity* of λ to small changes in transition probabilities between different stages in the life-cycle can be calculated from a matrix model.

4

The regulation of plant populations

Plant populations are not dusty museums of plant life where the same faithful individuals are to be found on every visit, but show more the constant activity of a railway station; witnessing a never-ending flow of new arrivals and departures. The timetables of these plant arrivals and departures are fecundity schedules and life tables. In view of this constant flux of individuals through plant populations a remarkable observation emerges about the net outcome of all these changes: individuals come and go but population size often remains more or less constant.

Population regulation and density dependence

If the death-rates and birth-rates which determine a population's size were subject to random changes it would only be a matter of time before such a population became extinct. Random processes countenance extremes, oblivious of any catastrophic consequences. The first extreme increase in the death-rate caused by a grazing animal or by adverse climate would finish off a population for ever. This is no idle observation because it suggests, on theoretical grounds alone, that populations which have a high flux of individual members and which are stable in overall population size must be cushioned from the random occurrence of high mortality by the action of some specific process.

Populations may be cushioned from local catastrophe in two basic ways: 1. by the immigration of plants from areas not affected by the catastrophe; or 2. by changes in fecundity or mortality in the population itself, which compensate for the effects of the catastrophe. The second of these is discussed in this chapter.

A population which shows the operation of clear limits on population size is said to be *regulated*. Population regulation operates via *density-dependent* processes which alter the numerical impact of fecundity or mortality on population size as population density changes. A density-dependent mortality factor is one that relaxes as population density declines, and thereby slows or halts population decrease. When population density increases, a density-dependent mortality factor kills an increasing proportion of the population. An example is seen in the relationship between seedling survival and the original density of seeds in the Wisconsin

Fig. 4.1 Density-dependent processes in two plant populations: (a) mortality in a population of sugar maple establishing from seed (Hett 1971); (b) fecundity in experimentally manipulated natural populations of *Vulpia fasciculata*. (Watkinson and Harper 1978)

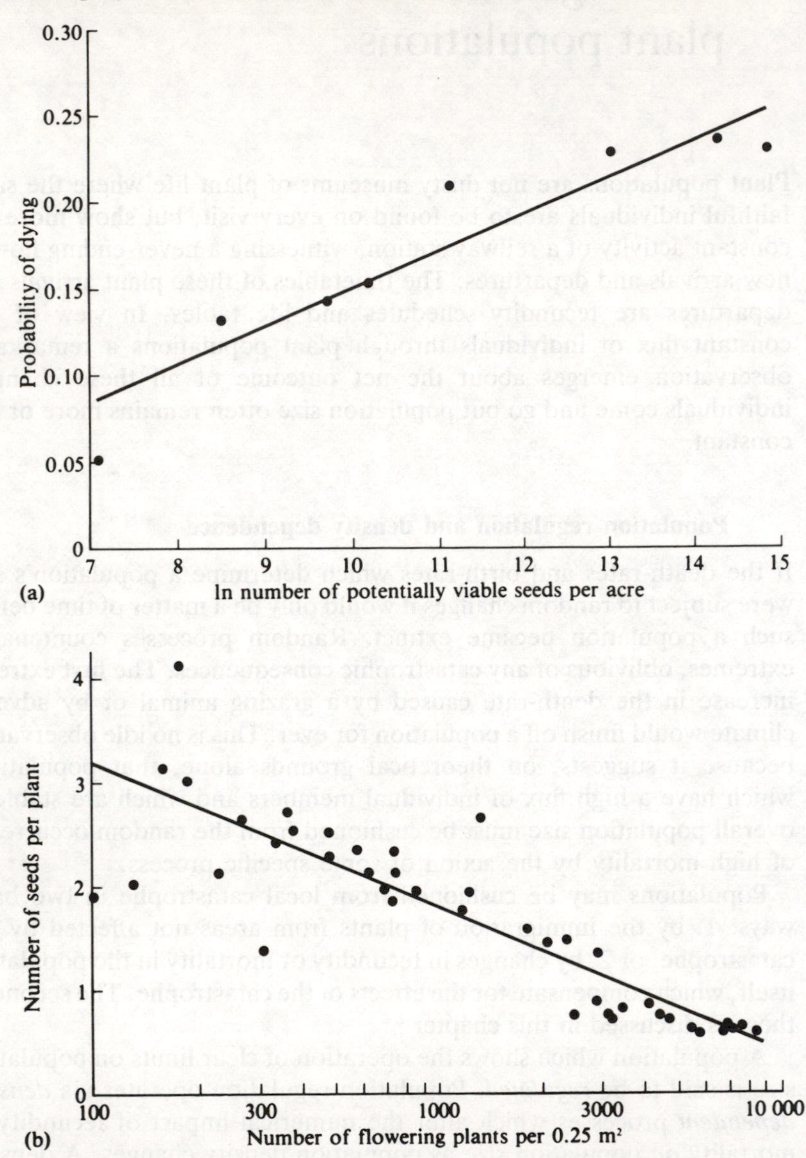

population of *Acer saccharum* studied by Hett (1971) (Fig. 4.1(a)). Density-dependent fecundity may also regulate population size by the production of fewer seeds per plant as population density rises (Fig. 4.1(b)).

The effect of all density-dependent processes is to produce populations of more or less constant density from populations which originally differ in size. Density-dependent mortality may reduce a wide range of seedling densities to a smaller range of adult densities, and density-dependent fecundity may result in similar numbers of seeds being produced by plant populations of widely different densities (Fig. 4.2(a)). The operation of both processes in some greenhouse populations of the grass *Bromus tectorum* planted at densities of 5, 50, 100 and 200 seeds per pot (Fig. 4.2(b)) illustrates how the combined effect of density-dependent mortality and density-dependent fecundity causes the population numbers to converge, so that approximately the same number of seeds begin the next generation, irrespective of their parent's planting densities (Palmblad 1968). Density-dependent mortality operates within a generation, reducing a large cohort of seedlings to a smaller cohort of adults. Density-dependent fecundity alters population size in a slightly less direct way.

Because density-dependent fecundity regulates the *seed* production of

Fig. 4.2 (a) An idealized diagram of density-dependent seedling mortality and density-dependent fecundity and their role in the regulation of a plant population. (b) An example of density-dependent mortality and density-dependent fecundity regulating population size in four experimental populations of an annual grass *Bromus tectorum*. (Data from Palmblad 1968)

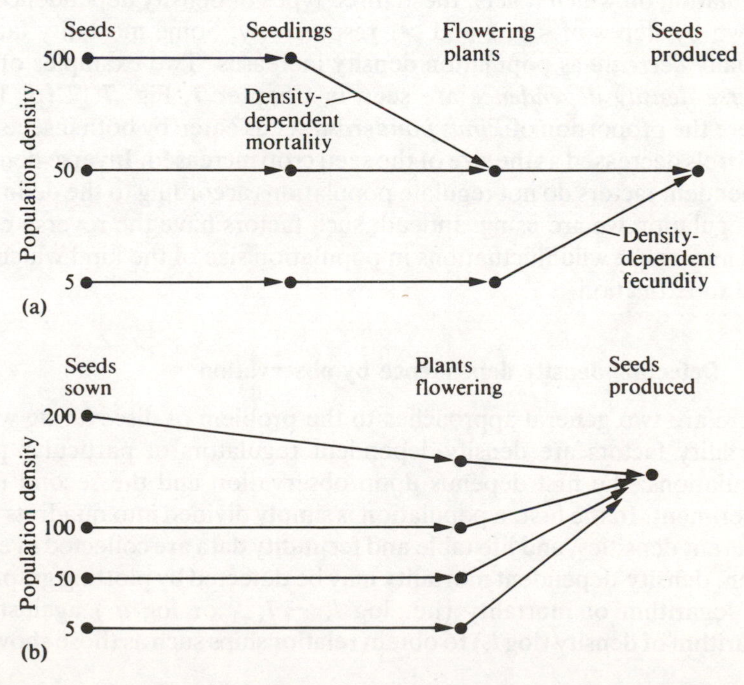

Fig. 4.3 Density-dependent mortality factors in a hypothetical population where they: (a) undercompensate for changes in population density; (b) exactly compensate for changes in population density; and (c) overcompensate for changes in population density. The diagonal (slope = 1) is drawn in as a dotted line.

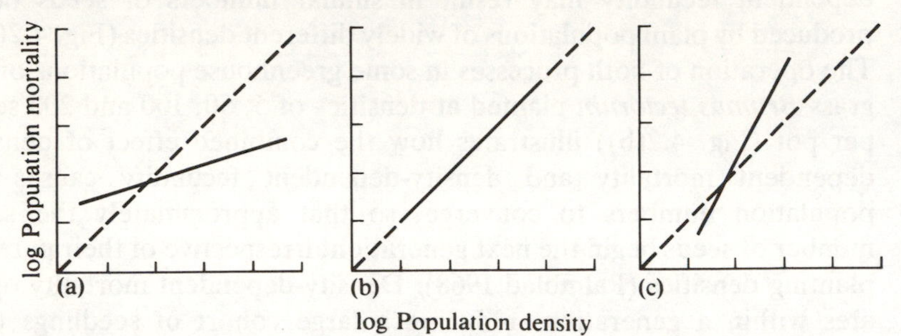

a population, it has an influence on the potential population size of the *next* generation of adults. In effect therefore, this is *delayed* density dependence: the effects of density upon mothers are visited upon their children.

Different density-dependent factors may regulate population size with varying effectiveness. Figure 4.3 illustrates a density-dependent mortality factor which *undercompensates* for population changes, one that *exactly compensates* for population changes and a factor that *overcompensates*. On a graph of mortality plotted against the density of the population on which it acts, these three types of density dependence are shown by slopes of <1, 1 and >1 respectively. Some mortality factors actually *decrease* as population density increases. Two examples of this *inverse density dependence* are seen in Chapter 7, Fig. 7.17 (p. 151), where the proportion of *Pinus ponderosa* seeds eaten by both insects and squirrels decreased as the size of the seed crop increased. Inverse density-dependent factors do not regulate populations according to the definition of regulation we are using. Indeed, such factors have the reverse effect and may cause wild fluctuations in population size of the kind which can lead to extinction.

Detecting density dependence by observation

There are two general approaches to the problem of discovering which mortality factors are density-dependent regulators of particular plant populations: the first depends upon observation and the second upon experiment. In the first, a population is simply divided into quadrats with different densities, and life table and fecundity data are collected in each. Then, density-dependent mortality may be detected by plotting graphs of the logarithm of mortality (i.e. $\log l_x - l_{x+1}$ or $\log q_x$) against the logarithm of density ($\log l_x$) to obtain relationships such as those shown in

Fig. 4.4 Density-dependent fecundity graphed as seeds per plot vs. density (a, b, c) and seeds per plant (d, e, f). (a) and (d) illustrate undercompensation; (b) and (e) exact compensation; and (c) and (f) overcompensation.

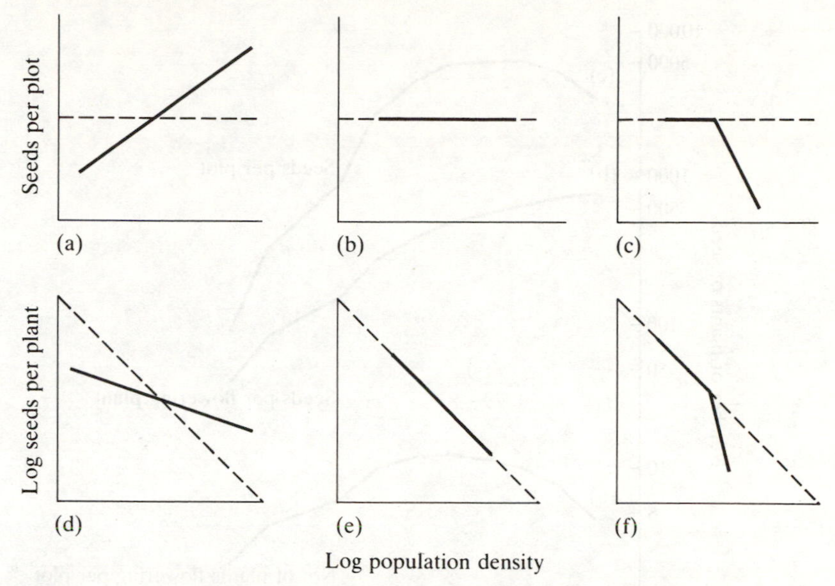

Fig. 4.3. Density-dependent fecundity can be detected from analogous graphs whose interpretation is given in Fig. 4.4. These graphs are useful for analysing the effect of fecundity and of individual mortality factors but they deal with only one factor at a time (see Silvertown (1982, p. 113) for a more complex analysis). The overall relationship we are interested in is that between N_t and N_{t+1} which is determined by density-dependent and density-independent factors affecting mortality and fecundity, all acting together.

In Chapter 3 we have already used a method for combining density-dependent and density-independent processes. In Fig. 3.3. we plotted the relationship between seedling density of *Erophila verna* in year *t* and the number of seeds produced per plot. The shape of this curve is determined by differences in mortality and population size structure which developed between plots of different initial density as plants grew. Within a few days of emergence the size of new seedlings became density dependent; differences between seedlings within Symonides' 10 cm × 10 cm plots increased rapidly with time and became most pronounced in plots with higher seedling densities. A large proportion of individuals in high-density plots never flowered but died or remained 'juvenile' rosettes (Fig. 4.5(a)). The non-flowering survivors crowded those plants which did flower and these produced fewer seeds the more they were crowded (Fig. 4.5(b)) with the result that fecundity was density dependent and showed *overcompensation*

Fig. 4.5 Density-dependent fecundity in *Erophila verna*. Note the logarithmic scale. (a) Number of plants flowering per plot; (b) seeds per flowering plant; and (c) seeds per plot, which is the logarithmic sum of curves (a) and (b). (Drawn from data in Symonides 1983a, b)

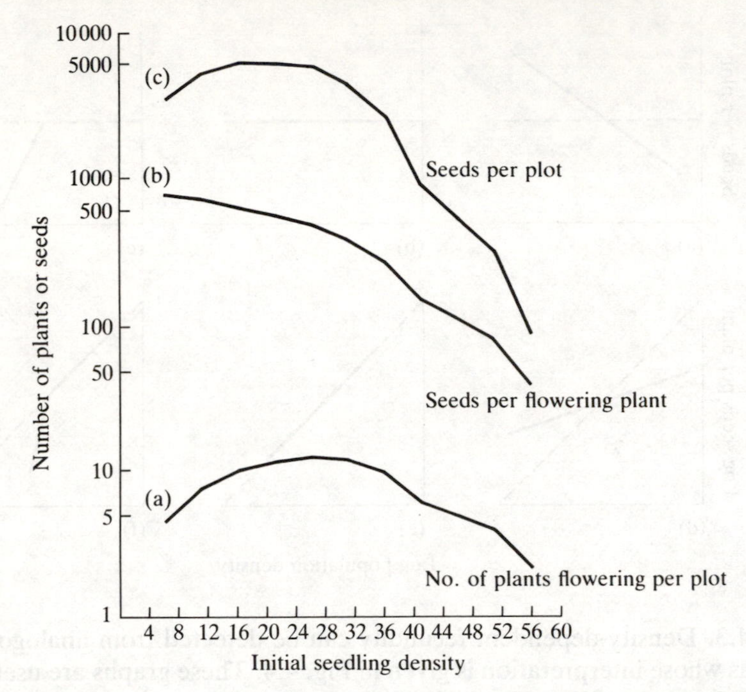

when seedlings exceeded 20 per 0.001 m^2 (compare Fig. 4.5(c) with Fig. 4.4(c)). This overcompensation produces the humped curve in Fig. 4.5(c) and Fig. 3.3.

Germination success was relatively independent of density and was therefore incorporated into the model by adding a straight line whose slope (N_{t+1} / N_t) is equal to the percentage germination of seeds. The germination line allows us to multiply every value of seeds produced by the N_t plants by a density-independent constant to obtain N_{t+1} seedlings. By varying the germination rate, which is a *density-independent* factor, we find that the *E. verna* population may exhibit stable (Fig. 3.3(a)) or cyclic behaviour (Fig. 3.3(b)).

Symonides found that the abundance of *E. verna* in many of her 10 cm × 10 cm plots cycled with a 2-year period, from 2–3 seedlings per plot in one year to over 50 per plot in the next (Symonides 1984). Cycling plots were not in phase with each other with the result that some overall stability of numbers was present on the larger scale when high and low density plots were averaged. Some of Symonides' 10 cm × 10 cm plots did not cycle in abundance at all, perhaps because germination success was consistently low in these (Fig. 3.3(a)).

This study of *Erophila verna* demonstrates the importance of spatial heterogeneity in plant populations. Plants compete with their neighbours and it is this neighbourhood competition which is responsible for density-dependent effects (Antonovics and Levin 1980). These effects can go undetected if population density is averaged over too large an area. For a plant as small as *E. verna*, even a 10 cm × 10 cm plot is quite a large one. The ultimate refinement which can be used to make sure that neighbourhood effects are properly accounted for, is to measure the size and distance of neighbours for every plant in a population and to use these data to find the influence of neighbours on each other and to predict each individual's fecundity and survival. This highly reductionist approach has been advocated by Pacala and Silander (Pacala and Silander 1985; Silander and Pacala 1985) but much useful information can be obtained by easier methods discussed here.

Observational methods such as those Symonides has used to such great effect will not always detect density dependence, even if it is present. If density dependence compensates nearly exactly there will be little natural variation in numbers of plants from place to place or from year to year and it will be impossible to use graphs such as Figs. 4.4 and 4.5 to reveal what is happening. It is paradoxical but true that density dependence is most difficult to detect in those populations which are most strongly regulated. The way round this is to manipulate density artificially.

Detecting density dependence by experiment

Watkinson and Harper (1978) obtained the range of *Vulpia* population densities in Fig. 4.1(b), p. 52 by adding seeds to some populations and thinning others by removing seedlings. In this case varying densities had no effect on mortality but altered individual plant size with the resulting effect on fecundity shown in the figure.

An experimental study of population regulation which elegantly demonstrates how the processes important in determining abundance may vary on a very local scale was carried out by Keddy (1981, 1982) on sea rocket *Cakile edentula*, an annual plant growing on a sand dune in Nova Scotia.

There were three sites on the seaward side of the dune (the top of the beach), in the middle, and on the landward side of the dune. Plots were sown with seeds at a range of densities at each site and survival and fecundity were recorded in each. The results are shown in Fig. 4.6 as graphs of mortality and fecundity against density for each site. For each variable against density the graph tells us two things: 1. the slope of the graph gives the strength of density dependence (a horizontal line, of 0 slope, = density independence); 2. the intercept on the vertical axis gives the level of density-independent mortality or fecundity.

Plants were large at the seaward end where there was no competing vegetation and where there was decaying eelgrass that supplied nitrogen.

Fig. 4.6 The relationship of fecundity and mortality to density in experimental plots sown with *Cakile edentula* at three sites on a sand dune. Statistically significant density-dependent relationships are shown by a regression line. There is no significant difference in levels of density-independent mortality at the three sites but there is a significant difference in the level of density-independent fecundity. The size and fecundity of solitary plants is far greater on the beach than elsewhere. (From Keddy 1981)

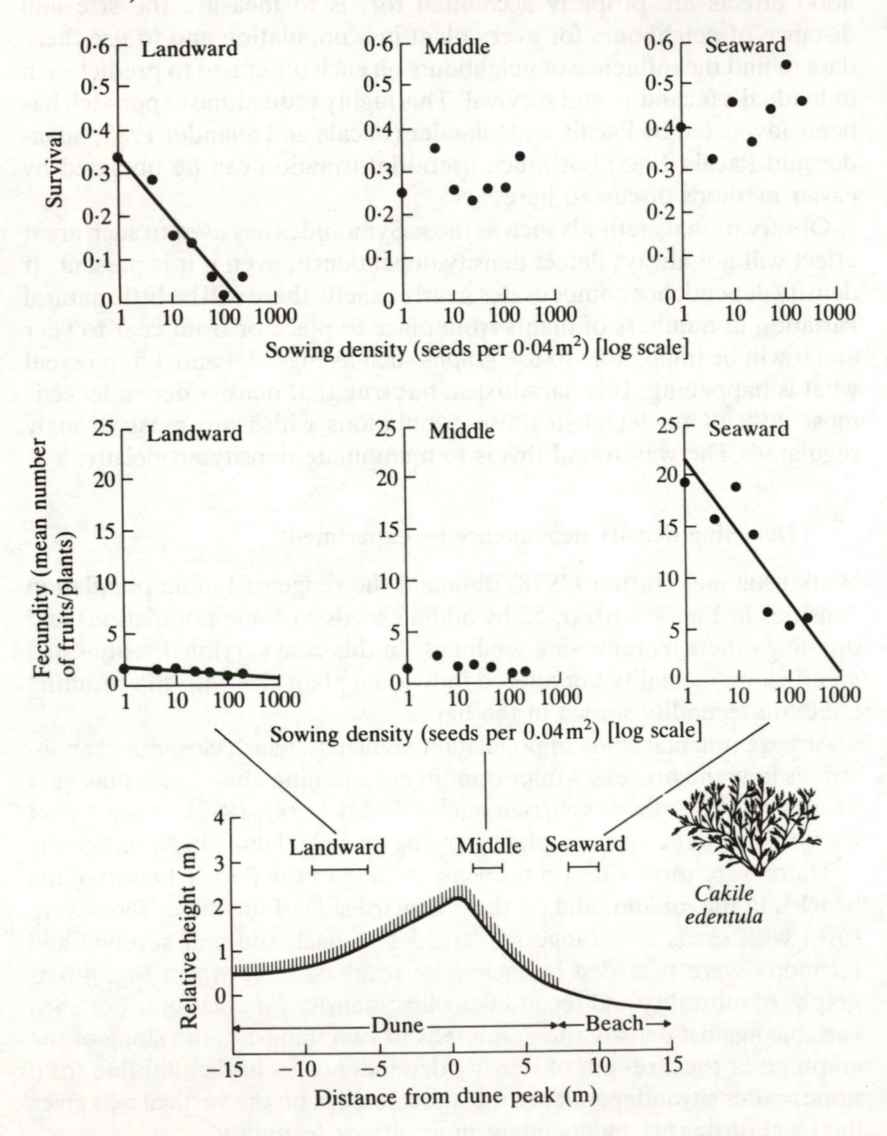

Here, increasing density resulted in a plastic adjustment of plant size and fecundity without causing mortality. Plants were small at the landward end and many of them were too small to reproduce. At this site, increasing density did not alter fecundity but did increase mortality. In summary; density in this study had different effects, depending upon the condition of plants, as determined by density-independent factors (nitrogen supply and competing vegetation) that influenced plant size.

Looking at Fig. 4.6 one would expect the natural density of *Cakile* to be greatest at the seaward end of the gradient. Mortality exceeded fecundity at the middle and landward sites so that one would expect the populations there to go extinct. In fact plants were most abundant in the mid-dune area because there was considerable migration of seeds inshore, taken there by tide and wind. Watkinson (1985) estimated that between 50 and 80 per cent of seeds produced in the beach population must have been carried landward to sustain the observed density of plants in mid-dune.

A word of clarification about population regulation is in order here. Density-dependent regulation is not a magic wand that can stop the advance of large perennials into the open areas where *Erophila verna*, *Cakile edentula* or *Vulpia fasciculata* grow, any more than it can stop a steamroller. It is the fate of plant populations on sand dunes to go locally extinct as succession proceeds and the survival of these plants in their habitats at large is dependent upon dispersal.

Self-thinning and the −3/2 power law

The plastic growth of plants, which allows 30-year-old trees to exist as dwarfed individuals beneath the canopy of giants not much older than themselves, and which causes *Erophila* and *Vulpia* to vary their seed production with population density, necessitates a special consideration of size when we try to arrive at some idea of how many plants can occupy a given space in a habitat.

If we imagine habitat space to be like a box capable of containing a given volume of children's wooden building blocks, it is obvious that such a box can accommodate a small number of large blocks or a large number of small blocks. If a number of small blocks are removed they can be replaced by a larger block of equivalent total volume: the *size* of wooden blocks and the *number* of wooden blocks are inversely related. In principle this rule also governs the number and size of plants that can exist in a population when closely packed.

While both trees and blocks obey the same rules of geometry, the analogy ends as soon as we take into account the fact that plants grow. Were plants to behave exactly like wooden blocks, a graph of log mean plant weight vs. log plant density for a full habitat 'box' would have a slope of −1 (Fig. 4.7). In fact this precisely inverse relationship between the size and density of plants does apply to populations of low density

Fig. 4.7 The relationship between mean plant weight and plant density expected when total plant biomass is at a maximum.

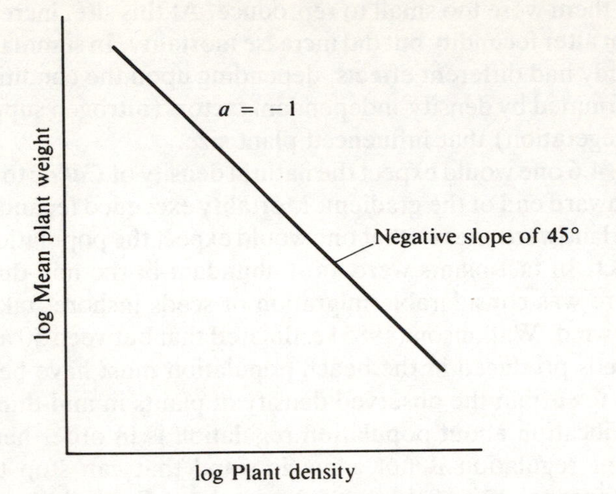

which have grown to such a weight that they have reached the carrying capacity of the environment. Or, in other words, populations which have filled the habitat box to its capacity for that species.

At lower plant sizes a different relationship exists between log mean plant weight and log plant density. In populations at very low density, log mean plant weight will increase just as fast as the plants can grow and without any change in density occurring. This is obviously what will happen to the relationship between mean plant weight and density in the extreme case where only one plant is present (e.g. the sparse population in Fig. 4.8). Populations with different initial densities may finally achieve the same total weight because there is a plastic adjustment of individual plant size (e.g. Fig. 4.2(b)). Sparse populations cease to grow in *total* plant weight when they encounter the carrying capacity of the habitat, which is shown in Fig. 4.8 by a line of slope −1. Any increase in mean plant weight which occurs after this line is reached, takes place at the expense of an exactly proportionate decrease in density, i.e. at the expense of plant survival.

Populations of small plants at higher densities also increase in mean plant weight as they grow, but mortality occurs before the carrying capacity is reached and before the increase in the total weight of the plant population ceases. Dense populations which have reached a size at which mortality occurs demonstrate a relationship between log mean plant weight and log density which generally has a slope of −3/2 (e.g. the dense population in Fig. 4.8). This means that every change of 3 log units in mean plant weight corresponds to a change of only 2 log units in mean plant density.

As plants in a dense population become larger with age, the density

Fig. 4.8 The progress of a sparse and a dense tree population through time, illustrating the main features of the −3/2 thinning process.

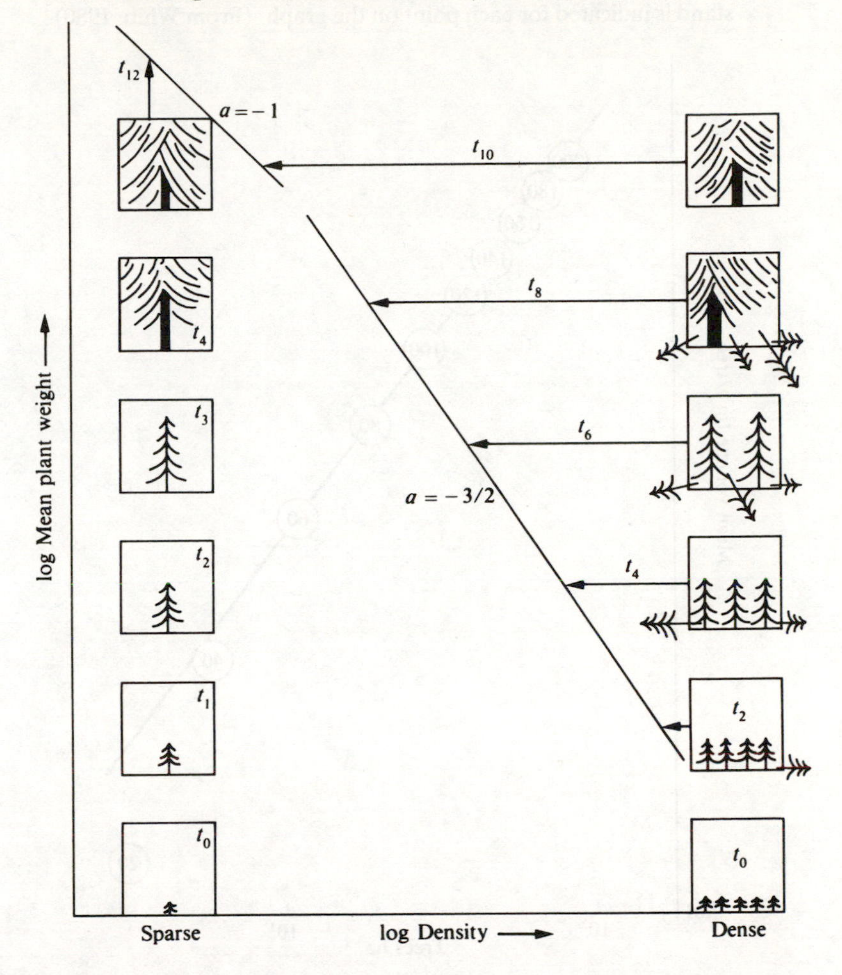

of individuals in the population decreases due to mortality. For as long as the relationship between mean plant weight and density is governed by a line with slope −3/2, total plant weight will increase because mean plant weight is increasing faster than density is falling. An example of this process, which is called *self-thinning*, is shown in Fig. 4.9. It has been observed quantitatively in about eighty species of trees and herbs whose populations show density-dependent mortality (White 1980). All demonstrate a relationship between the mean dry weight (w) of plants and population density (d) of the surviving individuals of the form:

$$w = Cd^{-a} \qquad [4.1]$$

where C and a are constants, a is the slope of the line in a log/log plot of

Fig. 4.9 The relationship between log mean volume per tree and log density of trees in ten stands of ponderosa pine. The age of each forest stand is indicated for each point on the graph. (From White 1980)

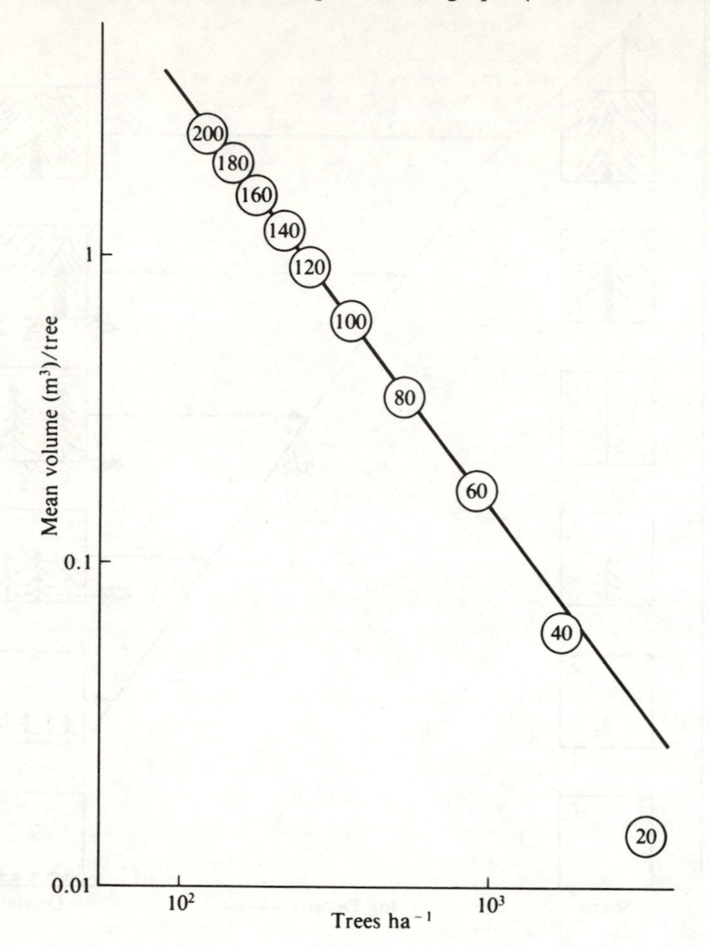

density and mean plant weight and log C is the intercept of the line on the ordinate:

$$\log w = \log C - a \log d \quad [4.2]$$

A further generalization about the self-thinning process which emerges from a comparison of different species is that C has a limited range of values between 3.5 and 4.3 (White 1980), and a frequently has a value in the region of $-3/2$ (Yoda *et al.* 1963).

The apparent constancy of the value of a led Yoda *et al.* to name [4.1] the *−3/2 power law*. To some extent the generality of the power law is an artifact of the way it is plotted on log–log axes and also the fact that *mean* plant weight mathematically must rise when mortality removes the smallest plants, even if the survivors do not grow at all. Both these factors tend to constrain the relationships that are possible between *w* and *d*. In a review of the power law Westoby (1984, p. 181) suggested that 'It may be best to speak of a thinning band rather than a thinning line.' A diagrammatic summary of the self-thinning process is shown in Fig. 4.8. Its main features are:

1. Plants increase in mean size (weight) with time.
2. No density-dependent mortality occurs until populations reach the self-thinning line.
3. Mortality begins earlier in dense populations than in sparse ones.
4. Trees of the same mean weight are younger in the sparse population than in the dense one (indicated by relative *t* subscripts).
5. Both sparse and dense populations eventually reach a stage where weight increments and mortality are balanced, $a = -1$ and total plant weight no longer increases.
6. The point where $a = -1$, is reached by denser populations first.

In the space beneath the $-3/2$ thinning line is a whole set of combinations of (relatively) low plant densities and mean biomass per plant, each of which may correspond to the situation prevailing in a population at a particular moment before density-dependent mortality begins to occur. When ryegrass (*Lolium perenne*) was planted at densities of 320–10 000 plants m^{-2} by Kays and Harper (1974), they first observed genet mortality in the densest populations, followed successively by plants in less dense treatments as each increased the mean weight per genet to the point at which it intersected the $-3/2$ thinning line (Fig. 4.10a). Because thinning began first in dense populations, these experienced greater overall mortality than less dense ones.

Kays and Harper also recorded changes in tiller number in the experiment and found that there was density-dependent tiller production by the surviving genets. At high densities the rate of genet mortality exceeded the rate of tiller formation and the overall tiller density fell. At low densities genet mortality was low and tiller formation high so that tiller densities rose. As a result of these two processes, density-dependent mortality reduced the 30-fold range of genet densities that were sown to a 6-fold range, and density-dependent tiller production caused all sowings to converge on the same tiller density by the end of the experiment (Fig. 4.10(b)). The $-3/2$ thinning line places an upper limit on the mean weight of plants in populations of a particular density, but exactly when this constraint acts depends upon how fast plants are growing and upon the initial density of the population. When the line is reached it thereafter determines just how much thinning must occur for a given increment in plant weight.

Various factors determine how fast self-thinning progresses, once it has begun. In a study of self-thinning in fleabane (*Erigeron canadensis*),

Fig. 4.10a Self-thinning in four populations of *Lolium perenne* planted at four different densities. H1–H5 are replicates harvested at five successive intervals. (From Kays and Harper 1974)

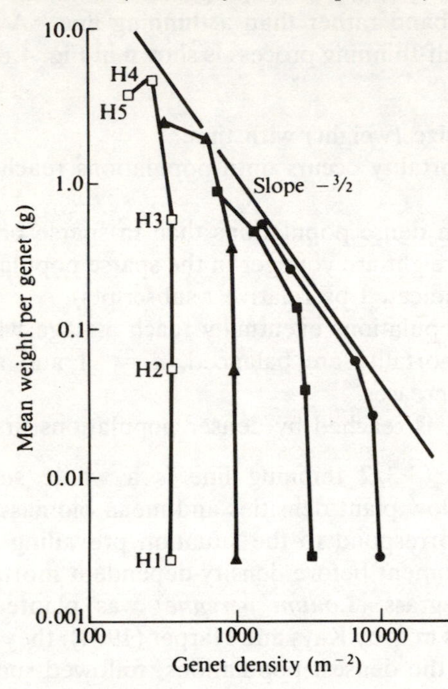

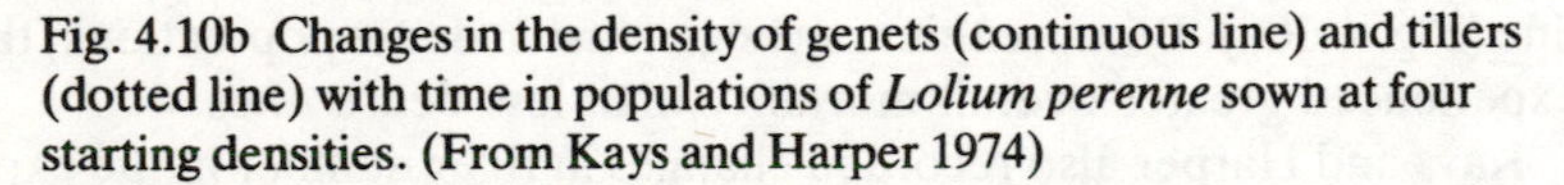
Fig. 4.10b Changes in the density of genets (continuous line) and tillers (dotted line) with time in populations of *Lolium perenne* sown at four starting densities. (From Kays and Harper 1974)

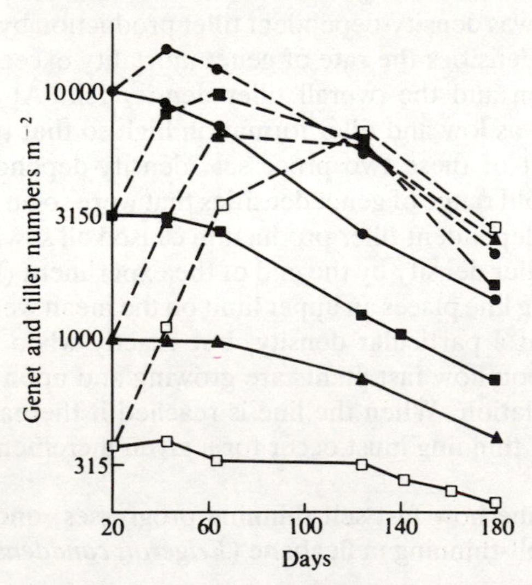

an annual growing in an abandoned field at Osaka in Japan, Yoda and his associates observed a pure population decline from 122 400 plants m^{-2} to 1060 m^{-2} in 9 months. Five plots in the field were treated with quantities of fertilizer applied in the ratio 5:4:3:2:1. The process of thinning was fastest in the most fertile plots which increased mean weight per plant more rapidly than populations in less fertile plots as they raced ahead of them up the thinning line.

Although $-3/2$ is apparently a common value for the slope of the thinning curve (a) it is not universal. Kays and Harper (1974) found that when they replicated their *Lolium* experiment in 30 per cent of full daylight, a acquired a value of -1, indicating that the balance between growth increments and genet mortality occurred at smaller plant size than in populations of the same density grown under full daylight.

A similar result was obtained by Hiroi and Monsi (1966) for sunflower (*Helianthus annuus*) grown in reduced light intensity. There are conflicting opinions in the literature (Westoby 1984), but the most plausible interpretation is that in low light (Lonsdale and Watkinson 1982) or on very poor soil (Morris and Myerscough 1984) the carrying capacity is lowered and the slope changes from $-3/2$ to -1 at a lower mean weight and higher (= earlier) density.

Although Yoda *et al.* (1963) originally derived the $-3/2$ power law empirically, they offered a simple explanation of it derived from the ratio of plant volume to the ground area a plant occupies, and hence the resources including light available to it. To follow their explanation we will again substitute wooden cubes for real plants for the sake of simplicity. The weight w of such a plant is proportional to its volume, which is the cube of its linear dimension l:

$$w \propto l^3 \qquad [4.3]$$

The mean area s occupied by the plant is proportional to the square of this linear dimension:

$$s \propto l^2 \qquad [4.4]$$

and therefore:

$$\sqrt{s} \propto l \propto \sqrt[3]{w} \qquad [4.5]$$

and $$w \propto s^{3/2} \qquad [4.6]$$

Self-thinning does not occur until plant density is high enough to produce 100 per cent cover. When this is reached, the mean area a plant occupies will be inversely proportional to density (d):

$$s \propto 1/d \qquad [4.7]$$

Substituting $1/d$ for s in [4.6] we get:

$$w \propto 1/d^{3/2} \quad \text{or} \quad w \propto d^{-3/2} \qquad [4.8]$$

and inserting a constant:

$$w = Cd^{-3/2} \qquad [4.9]$$

The other major factor determining the nature of the relationship between plant size and number is seen most clearly in the between-shoot interactions in clonal plants. When such plants are considered as genets they conform to the thinning law (e.g. *Lolium* in Fig. 4.10, p. 64), but when the dynamics of the component parts are analysed, the individual shoots rarely reach sufficient density for self-thinning to operate.

Self-thinning between genets is the result of crowding, over which individual plants have no control. By contrast, the density of connected shoots produced by a single genet is dependent upon the behaviour of an individual which may control the density of ramets it produces and their disposition, by the rate of growth and geometry of a rhizome. A genet which produced ramets at such a high density that self-thinning removed some of them would presumably be at a disadvantage compared to genets which did not crowd themselves in this way. Where shoots in a clonal population do occasionally reach a self-thinning density, the evidence suggests that the −3/2 thinning line is not transgressed (Hutchings 1979).

The mechanics of self-thinning

While the self-thinning of dense populations proceeds in a stately fashion and with measured step up the self-thinning line, less well-understood events are taking place between individuals within such thinning populations. The reciprocal relationship between mean plant weight and the density of surviving plants has often been observed, but from the point of view of the individual plant in such a population, its progress may be marked by death or survival and a varying amount of growth. What factors determine whether a particular plant lives or dies, grows rapidly or lingers on the verge of survival?

Some idea of what happens to individual plants during the self-thinning process can be obtained from changes in the frequency distribution of individual weight with time. In the case of marigold (*Tagetes patula*) (Fig. 4.11) sown at 400 plants to a 62 cm × 62 cm box, seedlings 2 weeks old had an approximately normal distribution of weight, with some under-representation of very small plants (Fig. 4.11) (Ford 1975). The larger seedlings at this stage were distributed at random, possibly owing their extra size to faster emergence from the soil and locally favourable microtopographical conditions of the kind discussed in Chapter 2. At 4 weeks (Fig. 4.11(b)), an upper canopy began to form from the leaves of the tallest plants. These were also the heaviest individuals and were evenly distributed in the population. Though very few plants had died at this stage, a frequency distribution strongly biased

Fig. 4.11 The changing pattern of individual plant weight in a population of marigolds, shown as the frequency of plants in twelve weight classes. Weight classes were determined by dividing the range of weights into twelve equal intervals. Size distribution at: (a) 2 weeks; (b) 4 weeks; (c) 6 weeks; (d) 8 weeks. (From Ford 1975)

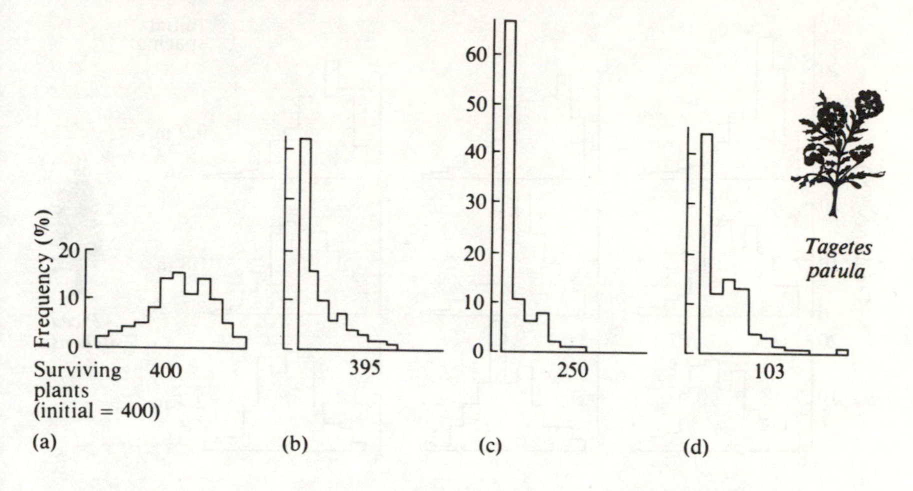

towards small plants (i.e. positively skewed) had developed. Self-thinning began between 4 and 6 weeks (Fig. 4.11(c)) and selectively removed the smallest plants. Between 6 and 8 weeks (Fig. 4.11(d) deaths occurred in a pattern which left survivors evenly distributed in space.

A similar sequence of events has been observed in the growth of a number of crop plants both in single species stands and when grown in mixtures, and also in tree populations (Ford 1975; Mohler, Marks and Sprugel 1978). These events have been described as the establishment of a 'hierarchy of resource exploitation, which results in differential growth rates among its members' (White and Harper 1970). Plants at the bottom of this hierarchy are referred to as 'suppressed', those at the top as 'dominant'. Self-thinning in pure stands of sitka spruce (*Picea sitchensis*) planted at different initial spacings shows that increased density accelerates the skewing of the size frequency distribution with time so that the hierarchy of exploitation is established more rapidly in dense stands (Fig. 4.12). In mixtures of species such as rape (*Brassica napus*) and radish (*Raphanus sativus*) (White and Harper 1970), and mustard (*Sinapsis alba*) and cress (*Lepidium sativum*) (Bazzaz and Harper 1976), resources are usually unequally divided between species so that one of them is over-represented in the suppressed class and the other over-represented among the dominants.

Among even-aged stands of trees, which generally follow the same course described for the early growth of *Tagetes*, the smallest trees may eventually be eliminated altogether, to give rise to a symmetrical

Fig. 4.12 The development of plant size distributions in populations of *Picea sitchensis* sown at three densities. Plant size is measured in twelve equal intervals of tree girth. (From Ford 1975)

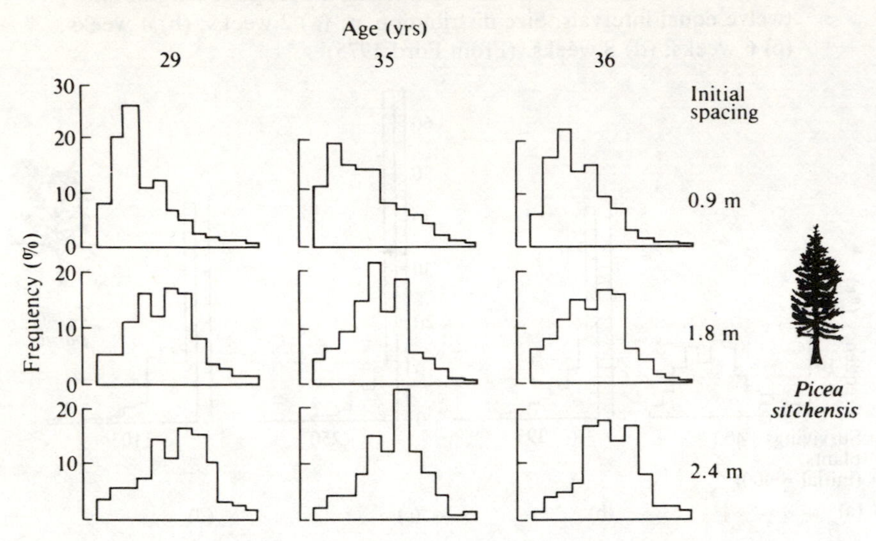

frequency distribution of sizes once again. An example of this is seen in a series of even-aged natural populations of balsam fir (*Abies balsamea*) (Fig. 4.13).

There are several ways in which skewed weight distributions which give rise to the 'hierarchy of exploitation' in dense populations may be produced. The first explanation is contained within the terms 'suppressed' and 'dominant' used to describe the plants at either end of the hierarchy. The use of these terms implies that large plants actively interfere with smaller ones through some direct competitive mechanism. To test this idea Turner and Rabinowitz (1983) grew the grass *Festuca paradoxa* in dense stands in flats and in isolated pots. The weight distribution of plants grown alone skewed first and remained more skewed throughout the experiment which lasted 44 days. Turner and Rabinowitz suggested that this result showed that skewing was the result of a particular relationship between the relative growth rate of a plant and its size. A little detailed argument is needed to explain this.

We have already seen that populations of plants such as *Tagetes* begin growth in dense stands with a normal frequency distribution of plant weight. Growth rate, measured in grams per plant per day, is directly proportional to plant weight at this stage so we may assume that it too has a normal frequency distribution. When growth takes place, the additional weight produced leads to an increase in growth rate. This increase in rate accelerates plant growth, the extra growth increases the rate and so on in an exponential fashion. This process of exponential growth does not go on indefinitely of course, but it may cause plants with initially only small

Fig. 4.13 Frequency distributions of trunk diameter for *Abies balsamea* stands divided into twelve equal intervals. Stands densities (d) are in units of trees m^{-2}. (From Mohler, Marks and Sprugel 1978)

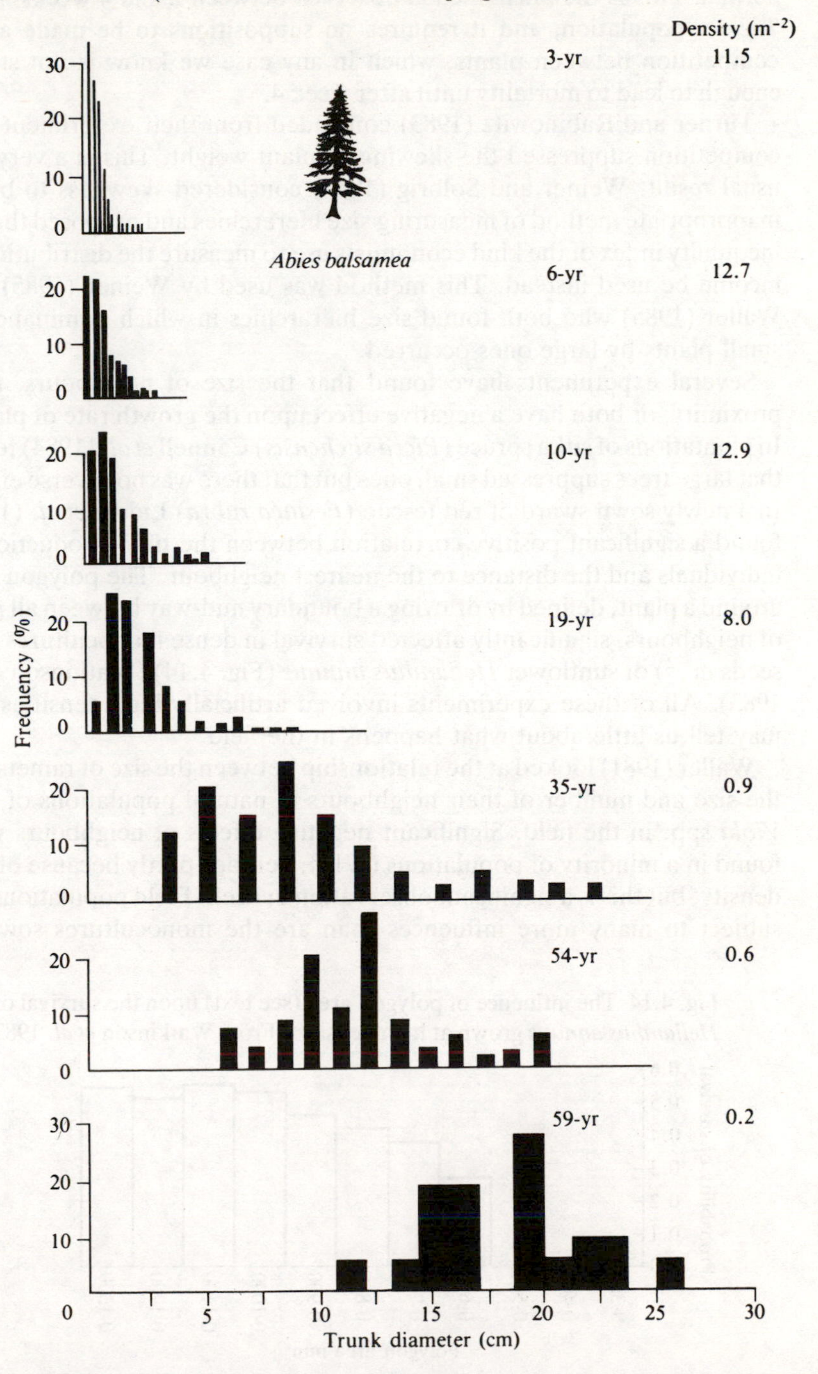

differences in weight to diverge very rapidly in size, with the consequence that a normal distribution of plant weight rapidly skews and becomes log normal. This is the phenomenon observed between 2 and 4 weeks in the *Tagetes* population, and it requires no suppositions to be made about competition between plants, which in any case we know is not strong enough to lead to mortality until after week 4.

Turner and Rabinowitz (1983) concluded from their experiment that competition suppressed the skewing of plant weight. This is a very unusual result. Weiner and Solbrig (1984) considered skewness to be an inappropriate method of measuring size hierarchies and proposed that an inequality index of the kind economists use to measure the distribution of income be used instead. This method was used by Weiner (1985) and Waller (1985) who both found size hierarchies in which dominance of small plants by large ones occurred.

Several experiments have found that the size of neighbours, their proximity, or both have a negative effect upon the growth rate of plants. In plantations of sitka spruce (*Picea sitchensis*) Cannell *et al.* (1984) found that large trees suppressed small ones but that there was no reverse effect. In a newly sown sward of red fescue (*Festuca rubra*) Liddle *et al.* (1982) found a significant positive correlation between the tiller production of individuals and the distance to the nearest neighbour. The polygon area around a plant, defined by drawing a boundary mid-way between all pairs of neighbours, significantly affected survival in dense monocultures (650 seeds m^{-2}) of sunflower *Helianthus annuus* (Fig. 4.14) (Watkinson *et al.* 1983). All of these experiments involved artificially high densities and may tell us little about what happens in the field.

Waller (1981) looked at the relationship between the size of ramets and the size and number of their neighbours in natural populations of four *Viola* spp. in the field. Significant negative effects of neighbours were found in a minority of populations (2/11), perhaps partly because of low density, but this is a significant observation in itself. Field populations are subject to many more influences than are the monocultures sown at

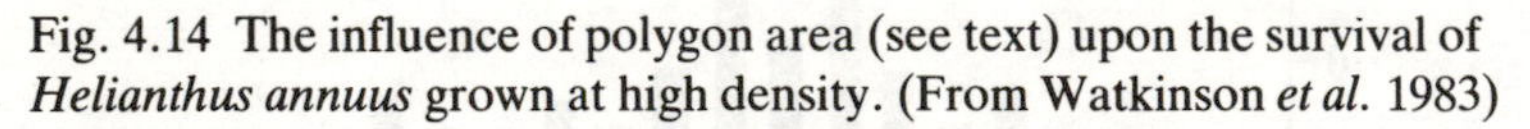

Fig. 4.14 The influence of polygon area (see text) upon the survival of *Helianthus annuus* grown at high density. (From Watkinson *et al.* 1983)

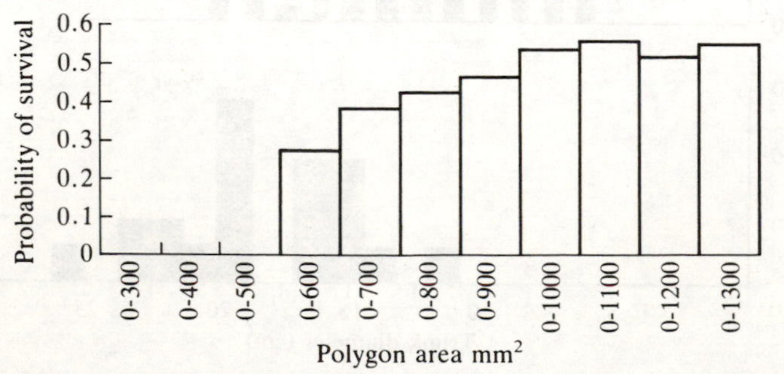

artificially high density in which experimenters try to eliminate (or just ignore) such things as microtopographical variation in the seedbed, differences in seed size and in time of seedling emergence and disease.

Hartgerink and Bazzaz (1984) sowed velvetleaf (*Abutilon theophrasti*) in heterogeneous and homogeneous seed beds at three densities. Aspects of heterogeneity explained 58–76 per cent of the variance in final size and seed output of plants. This is as large an effect as neighbours have in many experiments, but in this experiment there was no effect of neighbours on plant size. Benjamin (1982) did an experiment with cultivated carrot (*Daucus carota*) in which he controlled the degree of emergence synchrony and seed size variation and tested the effects of these variables upon the size of carrots grown at low ($25\,m^{-2}$) and high ($400\,m^{-2}$) densities. The mean weight of carrot roots and shoots was affected by density but not emergence synchrony, however *variation* in the weight of individual carrot roots at harvest was increased by both high density and asynchronous emergence. Variation in root size was not affected by seed size treatment. However, large or early-germinating seeds did have a competitive advantage and did produce larger plants than small or late-germinating ones.

In an experiment with two genotypes of skeleton weed (*Chondrilla juncea*), one resistant and one susceptible to a strain of the rust *Puccinia chondrillana*, the presence of disease determined which plants became dominant and which suppressed in high density 50:50 mixtures of the two (Fig. 4.15). Disease also changed the overall size–frequency distribution. The poorer performance of the resistant genotype in disease-free mixtures is an interesting manifestation of the 'cost of resistance' (Harlan 1976) which may explain why populations tend to loose resistance to a disease when no longer exposed to it.

Fig. 4.15 The frequency distribution of plant size in dense mixtures of two genotypes of *Chondrilla juncea* that were (a) disease-free; and (b) infected with a rust. Hatched columns are plants with a resistant genotype, stippled columns are susceptible plants and open columns are the sum of the two. (From Burdon *et al.* 1984)

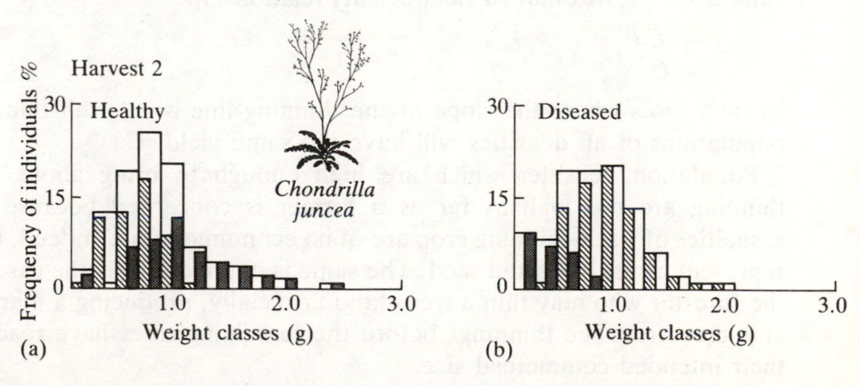

Population density and plant yield

From the point of view of an agronomist, the weight of plant material that can be obtained from a unit area planted at a given density may be of more interest than the relationship between mean plant weight and density which the $-3/2$ thinning law describes. If we were buying a box of wooden blocks that were sold by weight and we were interested in value for money, we would be more interested in the weight of blocks a box contained than in the mean size of those blocks. In our analogy the weight of blocks in a box is equivalent to the weight of plants in a unit area, or in other words the plant yield.

The relationship between yield and density in a thinning population is easily derived from the $-3/2$ thinning law. Since yield equals mean weight per plant times plant density, a yield corollary of the $-3/2$ thinning law which describes the relationship between yield and density in a thinning population can be derived as follows: The $-3/2$ law is:

$$w = Cd^{-3/2}$$

Yield = mean weight per plant × plant density:

$$y = w \times d$$

so:

$$w = y/d$$

Therefore by substitution:

$$y/d = Cd^{-3/2}$$

and cancelling both sides:

$$y = Cd^{-1/2}$$

For a population with a self-thinning line of slope $-3/2$ and intercept C this equation tells us the yield to be expected from a self-thinning population of density d. Yield/density equations for self-thinning populations with slopes shallower than $-3/2$ can be derived from the self-thinning law in the same way. For populations which have reached the upper part of the self-thinning curve where the slope achieves the value $a = -1$, we obtain a yield/density relationship:

$$\begin{aligned} y &= Cd^0 \\ &= C \end{aligned}$$

In such cases where the slope of the thinning line is -1, self-thinning populations of all densities will have the same yield, C.

Population densities which are high enough to bring about self-thinning are too high as far as a farmer is concerned because the casualties of a self-thinning crop are of no economic value. Indeed, they represent a cost in wasted seed. The same is not necessarily the case for the forester who may thin a tree stand artificially, producing a marketable crop from the thinnings before the remaining trees have reached their intended commercial size.

Two types of simple relationship between plant yield and population density have been observed for crop plants. A straightforward asymptotic relationship is found in various crops where the yield is measured in terms of whole plant weight or some vegetative part of this, such as the roots of a beet crop or the tubers of a potato crop (Fig. 4.16(a) and (b)). These two crops illustrate two further points of interest: 1. the yield of beet roots and beet tops respond in a parallel fashion to density; and 2. seasonal variation may affect the density at which maximum potato yield is reached. In 1956 planting densities higher than 10^4 parent tubers per acre achieved no higher yield of potatoes, but in 1958 a sowing of twice this density increased yield by about 30 per cent (Fig. 4.16(b)).

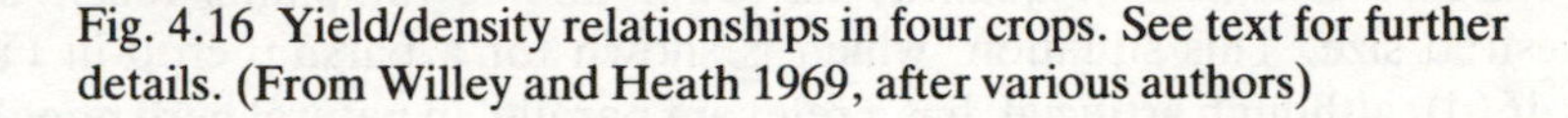

Fig. 4.16 Yield/density relationships in four crops. See text for further details. (From Willey and Heath 1969, after various authors)

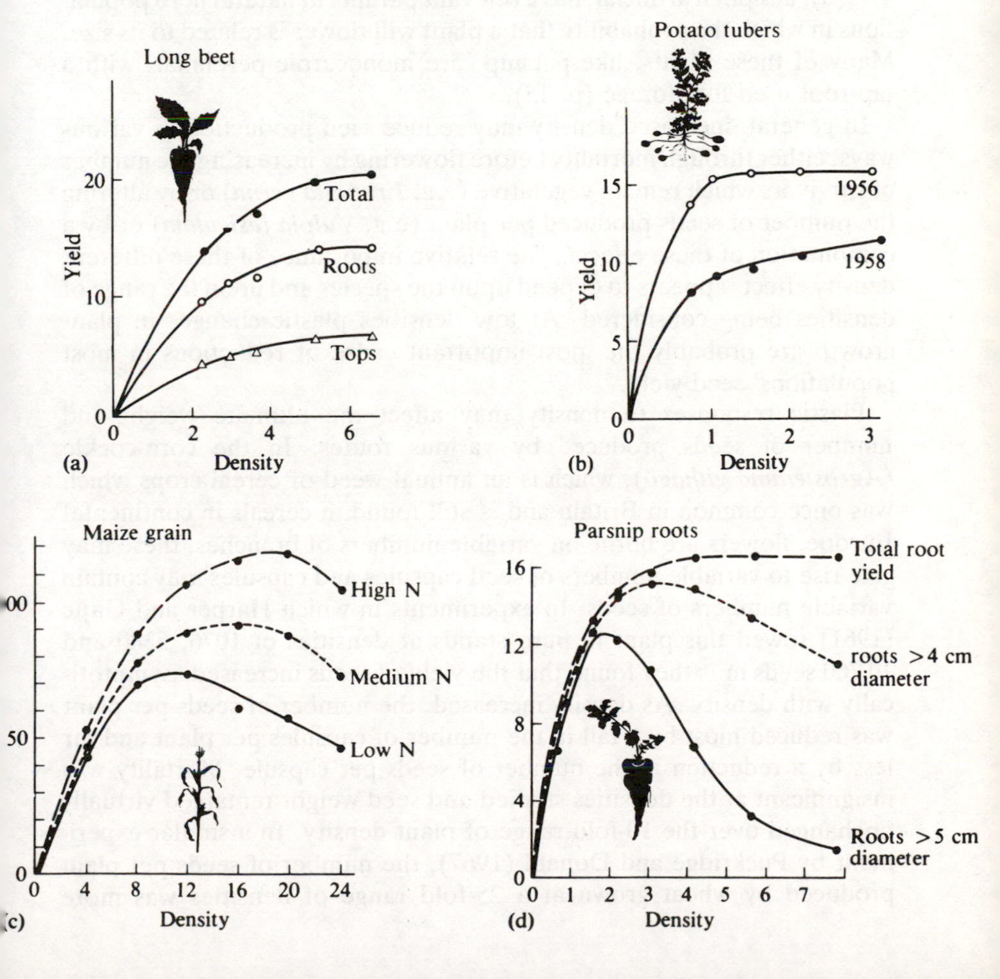

A second pattern of yield/density relationship is observed in crops grown for grain or seed. Grain or seed yield in crops such as maize (Fig. 4.16(c)), wheat, barley, soybeans, peas, ryegrass and subterranean clover grown for seed have all shown a parabola-shaped yield/density curve in experiments. In these crops, *total* plant weight is usually asymptotically related to density, but the yield from *reproductive parts* of the plant increases to a maximum and then actually decreases again as population density is raised above this. The density at which maximum seed yield is attained may vary with the nutrient status of the plants (Fig. 4.16(c)).

If the grower is interested in producing a graded crop of vegetables, only plants above a certain size may be regarded as contributing to yield. In this case even though total plant weight may be asymptotically related to density, the selected yield forms only a fraction of this and is bound to decrease in quantity as fewer and fewer plants reach the desired size. This situation, which is shown for a parsnip crop in Fig. 4.16(d), although artificial, has a relevant parallel in natural herb populations in which the probability that a plant will flower is related to its size. Many of these plants, like parsnip, are monocarpic perennials with a tap-root used for storage (p. 15).

In general, increased density may reduce seed production in various ways, either through mortality before flowering by increasing the number of survivors which remain vegetative (e.g. *Erophila verna*) or by altering the number of seeds produced per plant (e.g. *Vulpia fasiculata*) or by a combination of these effects. The relative importance of these different density effects appears to depend upon the species and upon the range of densities being considered. At low densities plastic changes in plant growth are probably the most important cause of reductions in most populations' seed yield.

Plastic responses to density may affect the ultimate weight and number of seeds produced by various routes. In the corn-cockle (*Agrostemma githago*), which is an annual weed of cereal crops which was once common in Britain and is still found in cereals in continental Europe, flowers are borne on variable numbers of branches, these may give rise to variable numbers of seed capsules and capsules may contain variable numbers of seeds. In experiments in which Harper and Gajic (1961) sowed this plant in pure stands at densities of 1076, 5380 and 10 760 seeds m^{-2} they found that the yield of seeds increased asymptotically with density. As density increased, the number of seeds per plant was reduced most by a fall in the number of capsules per plant and far less by a reduction in the number of seeds per capsule. Mortality was insignificant at the densities studied and seed weight remained virtually unchanged over the 10-fold range of plant density. In a similar experiment by Puckridge and Donald (1967), the number of seeds per plant produced by wheat grown at a 25-fold range of densities was more affected by a decrease with density in the number of ears per plant than by a decrease in the number of seeds per ear. The mean weight of wheat grains remained practically constant in this experiment too and no mortality occurred over the range of densities.

The manner in which reproductive output is adjusted with density depends upon the morphology of the plant. Both corn-cockle and wheat possess a structure of branches or tillers which can be multiplied during the growth and development of the plant as the resources available to it allow. When resources are scarce due to high density the first economy a plant makes is in these basic components of plant structure. The situation is different in another annual species, the cultivated sunflower (*Helianthus annuus*), which has but a single stem terminating in a single capitulum (a disc-like flower head). In some experiments by Clements, Weaver and Hanson (1929), this species was grown in a series of stands in which plants were spaced at intervals of 5, 10, 20, 40, 80, and 160 cm. Various dimensions of the plants were measured during the growth of the stands. Over the 80-fold range of densities, mean above-ground dry weight varied by three orders of magnitude (2 g/plant to 491 g/plant), leaf area varied by four orders of magnitude (41 cm^2 plant to 27 192 cm^2/plant), but mean plant height varied only 2-fold, from the plants spaced at 5 cm intervals which were 101 cm high to those 160 cm apart which were 218 cm.

In the early weeks of the experiment plants at higher densities were actually taller than those at lower densities. This early difference in height in response to density occurred during the period when the capitulum was developing from the apical meristem. The additional height growth among plants at high density contrasts strongly with the effect of density on seed production and seed weight, both of which were drastically reduced in these plants. Seed number per capitulum varied by two orders of magnitude (15 seeds/capitulum to 1803 seeds/capitulum) and the weight of individual seeds by one order of magnitude (0.009 g to 0.059 g). Having only a single capitulum, sunflower adjusts its seed production in the only way it can, by reducing seed number and by sacrificing seed weight. Plants from wild populations of *Helianthus annuus*, unlike their cultivated derivatives, are branched and respond to density by reducing the number of branches and the number of capitula (Harper 1977).

The adjustments plants make in their structure in response to density, discussed here mainly in terms of reproductive organs, nicely illustrate one of the consequences of modular construction in plants.

Summary

Population *regulation* operates via *density-dependent* processes of mortality and fecundity. When density dependence *compensates* exactly or *undercompensates*, these processes stabilize fluctuations in population numbers. *Overcompensation* can cause fluctuations in abundance. Density dependence can be detected by experiment or by observation, but the latter method may not work when regulation is strong. A combination of density-dependent regulation and *density-independent factors* determines local population abundance. Both may vary from site to site.

A reciprocal relationship exists between plant size and density. At low densities there are plastic responses of plant size as plants grow. At high densities, mortality occurs and plants tend to follow the −3/2 thinning law. At high mean plant weight, mortality and growth are balanced and no further increase in yield occurs.

As self-thinning progresses, the frequency distribution of individual plant weight skews and a hierarchy of exploitation is established. High initial plant density speeds up the skewing process. In mixed species stands, the species may be unequally represented among the dominant and suppressed plants in the hierarchy. Time of emergence, seed size, genotype, and heterogeneity in the seed bed may all affect which plants become largest in a crowded population.

A relationship between yield and density may be derived from the −3/2 power law. For whole plants, yield is usually asymptotically related to density, but the relationship for some plant parts shows there to be an optimum planting density to achieve the highest yield.

5

The demography of some plant populations

Demography is the corner-stone of a variety of studies designed to serve many different ends. The fundamental nature of the information contained in the life table and fecundity schedule means that it is of interest from many points of view, both theoretical and applied.

Conservationists and foresters practising some kinds of management are both interested in monitoring (and perhaps increasing) the rate of natural replacement which occurs in plant populations. Agronomists may require similar information to manage pastures and rangelands. Both may also be interested in the size of the fraction of a self-renewing population which may be harvested without harming it. Wildlife managers have spent considerable effort collecting information on the fruiting habits of wild shrubs and trees which provide food for animals.

Population ecologists interested in more theoretical questions about plant populations depend heavily upon the results of these applied studies. One of the functions of the population ecologist is to produce useful generalizations. Demography may help to do this in two areas. Firstly, it supplies information on the dynamics of populations, including the rate at which one individual is replaced by another. Secondly, where data are available on genetic differences between individuals, it may supply information on the relative replacement rate of different genotypes. Hence demography is also the corner-stone of studies in population genetics and evolution.

Two methods exist for obtaining the information required to fill out any life table with values of age-specific survivorship (l_x) and age-specific mortality (d_x). The first is simply to follow the demise of a cohort of seedlings, taking censuses at intervals. This is the only method available for analysing populations with non-overlapping generations (or cohorts) such as those of *Phlox drummondii* discussed earlier (Ch. 2, p. 10). A life table produced by this method is known as a *dynamic* or *horizontal* life table. It is often impractical to make a census of cohorts of long-lived organisms such as trees throughout their lifetime and a short-cut method must be used. In these cases values of l_x and d_x are calculated from the *age structure* of the population at a single (rarely at two) sampling date. Life tables produced by this method are known as *static* or *vertical* life tables. Assumptions have to be made about recruitment to populations analysed in this way and static life tables are consequently more prone to error. Their use is discussed separately in this chapter on p. 97.

The patterns of mortality observed in different plant populations appear to vary as a result of two things: firstly, species appear to have roughly characteristic patterns of mortality which depend upon whether they are annuals or perennials, herbs or trees. This is hardly surprising since these colloquial terms describe life history patterns which must be reflected to some extent in the life table. Secondly, their mortality patterns are modified by the particular circumstances prevailing in the different habitats which populations of the same species occupy. The density of a population is one important factor which affects mortality (Ch. 4).

Genetic variation within a species, interacting with local environmental conditions, may be responsible for significant demographic differences between populations. It is not yet clear how far we may generalize about the demography of a species, or even of a group of species such as annuals, even though they superficially share the same life history (but see Ch. 7). For convenience, this chapter is divided into sections dealing with different life-history groups, but the generalizations which are drawn from the demographic studies in each section are rather broad ones, many of which also apply to other plants. Clonal plants are considered in detail in Chapter 6.

Annuals and semelparous perennial herbs

These plants tend to be small, short-lived, flower only once and often occur locally in large numbers. The rosettes of some so-called semelparous perennials such as foxglove (*Digitalis purpurea*) (van Baalen and Prins 1983) and the ragwort *Senecio jacobaea* (see below) do often produce vegetative offshoots and, strictly, only the ramets not the genets of these plants are semelparous.

The death rate is given by the slope of the survivorship curve when l_x or N_x is plotted on a logarithmic axis. Virtually all of the annual populations whose seedling survivorship curves are depicted in Fig. 5.1 grow in habitats which are subject to annual or more frequent disturbance. *Alyssum* and *Leavenworthia* colonize bare soil, *Cerastium* is a plant of sand dunes and *Salicornia* occurs on tidal mud. The obvious hazards of these habitats do not appear to contribute significantly to the major episode of mortality in these populations which occurs after flowering. Two exceptions are *Sedum smallii* and *Minuarta uniflora*, both annuals growing on rock outcrops where seedlings are liable to be washed away by rain.

Note that the survivorship curves in Fig. 5.1 do not extend over the seed state of the life-cycle. Mortality during this stage accounted for 80 per cent of seeds produced by *Phlox* and *Sedum* and 94 per cent by *Minuarta*. Survivorship curves are a useful way of representing the over-

Fig. 5.1 Some representative survivorship curves for annual plants: (1) *Alyssum alyssoides* (Baskin and Baskin 1974); (2) *Cerastium atrovirens* (Mack 1976); (3) *Leavenworthia stylosa* (Baskin and Baskin 1972); (4) *Phlox drummondii* (Leverich and Levin 1979); (5) *Sedum smalii* and (7) *Minuarta uniflora* (Sharitz and McCormick 1975); (6) *Salicornia europaea* (Jeffries, Davy and Rudmik 1981).

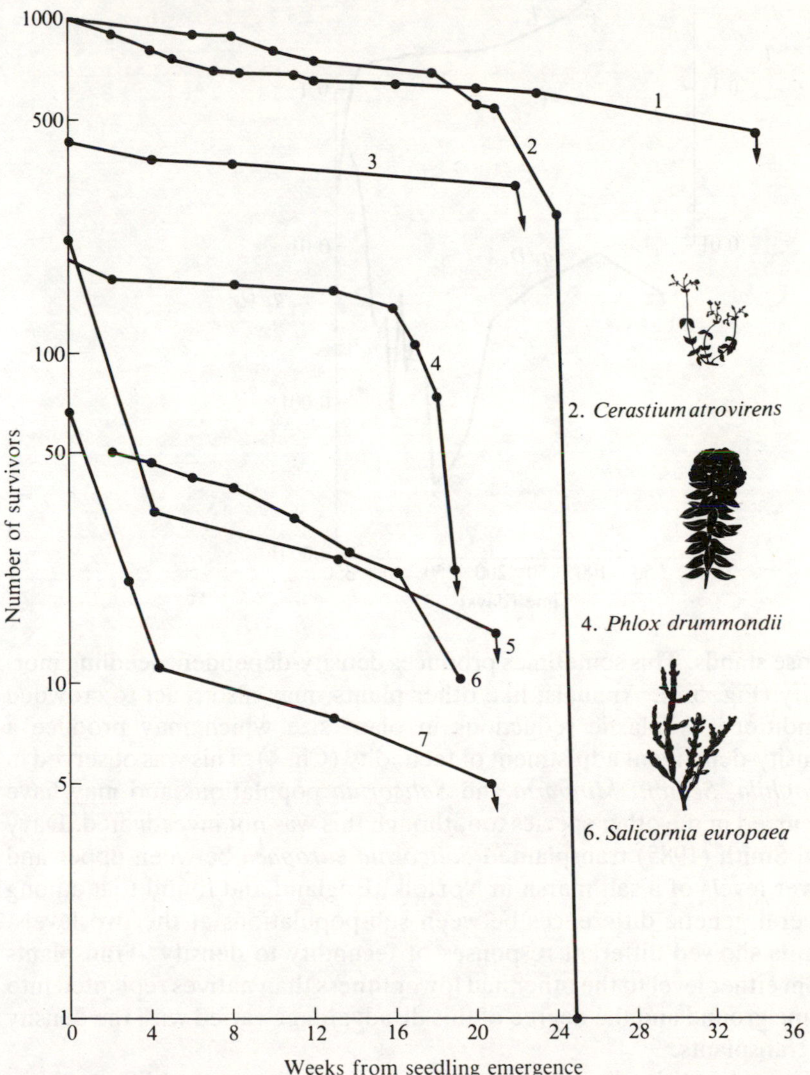

all pattern of mortality in a population, but these curves obscure detailed changes in mortality rate that occur from one census to the next, as Fig. 5.2 shows.

Because they colonize sites of local disturbance, annuals often occur in

Fig. 5.2 A survivorship curve (l_x) for *Phlox drummondii* and a graph of the daily mortality rate (q_x/D_x) for comparison. (Data from Leverich and Levin (1979) and Table 2.1(a))

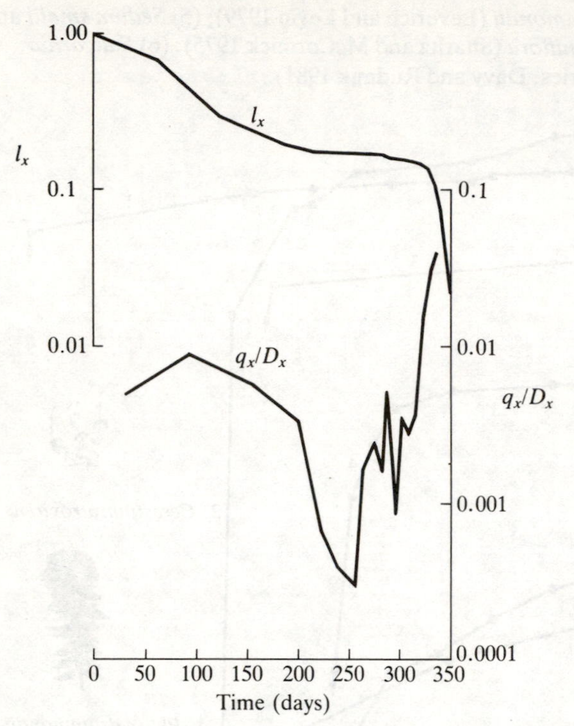

dense stands. This sometimes produces density-dependent seedling mortality (Fig. 5.3). Annuals, like other plants, may also react to crowded conditions by plastic reductions in plant size which may produce a density-dependent adjustment of fecundity (Ch. 4). This was observed in *Erophila*, *Sedum*, *Minuarta* and *Salicornia* populations and may have occurred in the other species too, though this was not investigated. Davy and Smith (1985) transplanted *Salicornia europaea* between upper and lower levels of a salt marsh in Norfolk, England and found that among several genetic differences between sub-populations at the two levels, plants showed different responses of fecundity to density. Transplants from either level to the other had lower fitness than natives replanted into home ground and the degree of this disadvantage varied with the density of transplants.

It can be misleading to contrast survivorship curves for different species because the shape of a curve reflects local conditions which can vary on a small scale. For example, there were significant differences in survival of poison hemlock (*Conium maculatum*) between ten quadrats censused by Tremlett *et al.* (1984), though all were within a few metres of each other. In this study, differences between quadrats were consistent

Fig. 5.3 Survivorship curves for natural populations of *Erophila verna* occurring in plots at densities per 0.001 m sq. of: (a) 1–2; (b) 5–10; (c) 15–30; (d) 35–50; (e) ≥ 55. (From Symonides 1983a)

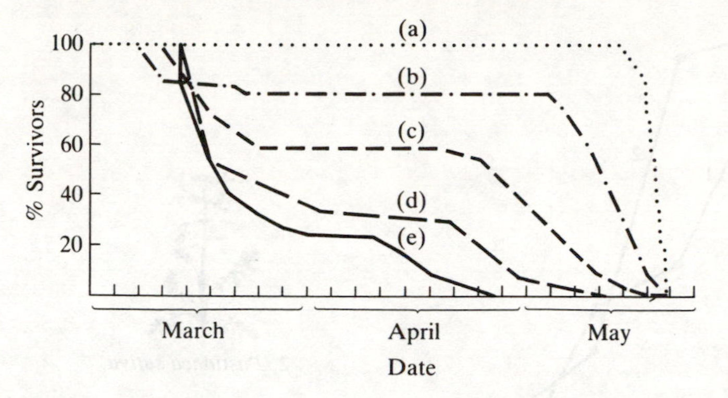

for different cohorts, even though the date of seedling emergence also affected survival.

We have already seen that emergence *order* can create demographic differences between cohorts (Ch. 2). Differences can also arise because mortality factors are *season*-specific. In a study of the annual grass *Bromus tectorum* (downy brome) which is an Eurasian plant that now occurs as a major constituent of the steppe grasslands in Washington State, Mack and Pyke (1984) found that some of the causes of mortality changed through the season. Germination of this species begins in the autumn. In some years desiccation was a serious cause of mortality, but only from September to November. Smut (*Ustilago bullata*) occurred only in June/July. Seedlings are most susceptible to both these causes of mortality and hence cohorts faced different hazards, depending upon when they appeared. The cohort of early-emerging seedlings tended to suffer from desiccation, the later ones suffered more from smut.

Survivorship curves for three semelparous perennial herbs are shown in Fig. 5.4. These plants are also colonizers of disturbed sites but delay seed production until the second or even later years of growth. Herbs behaving in this way are often described by the misnomer *biennial*. *Melilotus alba* is a true biennial, and all individuals in the population studied by Klemow and Raynal (1981) flowered or died by the second year. Individual plants of *Grindelia* and *Pastinaca* may delay flowering beyond their second season. Flowering in these species depends upon rosettes reaching a minimum size (see Ch. 2).

The delayed reproduction found in semelparous (monocarpic) perennials places constraints on growth and seed production which do not exist for annuals which complete their life cycle within 12 months. Whereas an annual may succeed in colonizing a gap and reproducing in it before it is closed by the invasion of surrounding perennial vegetation, perennial

Fig. 5.4 Survivorship curves of three monocarpic perennials: (1) *Grindelia lanceolata* (Baskin and Baskin 1979a); (2) *Pastinaca sativa* (Baskin and Baskin 1979b); (3) *Melilotus alba* (Klemow and Raynal 1981)

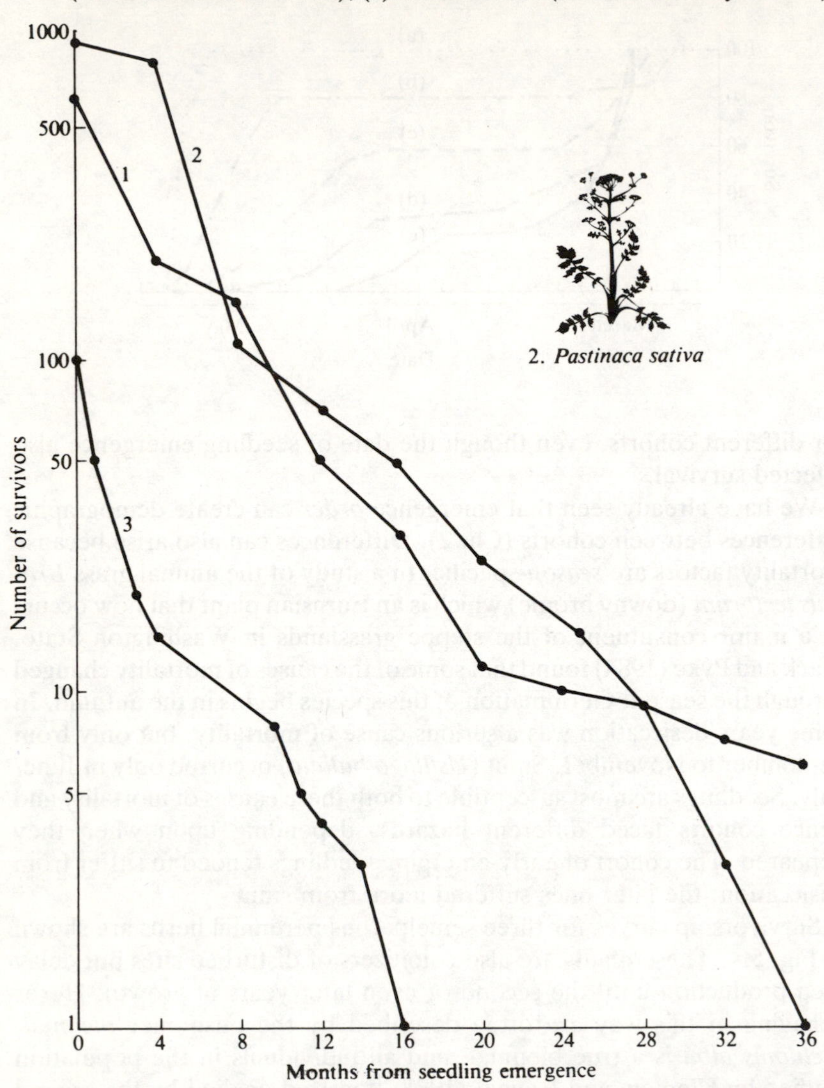

monocarps must contend with the interference of invading plants during two seasons of growth at least. Since most gaps begin to close very soon after they are created, the time a biennial arrives in such a gap is critical to its final success in producing seeds. This has been demonstrated for mullein (*Verbascum thapsus*) (Gross 1980) and wild carrot (*Daucus carota*) (Holt 1972) colonizing old fields in the USA. Both species may germinate in patches of bare soil where vegetation cover is reduced, but the probability of a rosette finally reaching sufficient age or size to flower is dependent upon how much interference is present from other vegetation. In carrot, mullein (Reinartz 1984) and other species there is variation within and between populations in the time of flowering. Within-population differences in carrot have a genetic component (Lacey 1986).

By the time a semelparous perennial does flower, the site which saw its birth has usually become closed to successful establishment by that plant's progeny. Harper (1977) has remarked that when biennials at a particular site attract the attention of an ecologist by their flowering, they are generally about to become locally extinct. Looking at this from the plants' point of view, one could say that the appearance of an ecologist in the area is an ill omen! *Chaerophyllum prescottii*, the semelparous perennial mentioned in Chapter 2 (p. 16), has overcome the problem of re-establishment by synchronizing seed production with events which generate new gaps. *Verbascum thapsus*, foxglove *Digitalis purpurea* (van Baalen 1982) and many other semelparous perennials have large seed pools and employ persistent seeds as a means of colonizing gaps as the opportunity arises.

Many semelparous umbellifers have neither a seed pool nor the persistent tuber of prescott chervil and are consequently confined to habitats in which regular disturbance permits regular re-establishment from the previous year's seed. Siberian hogweed (*Heracleum sibiricum*) is among several species of flood meadows near the river Oka in the USSR which depend upon this source of recruitment (Rabotnov 1978). The life cycle of these plants is something like a relay race. No period of inactivity in the soil is permitted between one generation and the next. The occasions when whole populations become extinct because disturbance fails to occur after flowering must be reduced to some extent by the perennial habit and the variable time plants of the same cohort may take to reach the flowering stage.

Iteroparous perennial herbs

This is a diverse group of plants which includes species that are sometimes semelparous (e.g. *Digitalis purpurea*) and many that are clonal. The latter are dealt with in Chapter 6. All species tend to have overlapping generations, though populations may be dominated by the founding individuals which first colonize a site. This is clearly illustrated by grey hair grass (*Corynephorus canescens*), a pioneer dune species that fixes mobile sand. At three sites in Poland where she studied this plant Symonides (1979a) found abundant seedlings each year, clumped around established individuals (tussocks) of the species. The seed distribution of this species in the soil was also clumped around tussocks (Symonides 1978). Figure 5.5(a) shows the distribution of seedlings and tussocks at a site newly colonized in 1968, the first year of the study. Seedling mortality was high during the year (Fig. 5.5(c)), particularly among seedlings emerging late relative to the rest of the cohort and those some distance from established tussocks.

Despite this mortality, some recruitment to the population of established plants did occur. Figure 5.5(b) shows how large annual flushes of seedlings produced annual increments in the established population, which eventually stabilized at a density of 170 tussocks in the 4 m^2 plot. The sequence of diagrams in Fig. 5.5(a) shows that recruitment in this population leads to gradual accretion of new individuals around the oldest tussocks.

As the process of accretion progressed, the age structure of established tussocks developed from a juvenile to a senescent form (Fig. 5.5(f)). This change in age structure occurred remarkably quickly compared to the average lifespan of an established plant which Symonides calculated to be about 7.8 years at this site. At two other sites where the colonization process had begun earlier and hence progressed further, senescence was further advanced, with a greater proportion of individuals in older age classes.

In the early stages of population growth, a high proportion of established individuals fruited (Fig. 5.5(d)). As population size increased, tussocks in younger age classes fruited less frequently than older ones because of the effects of crowding. By 1975 reproduction was practically confined to tussocks more than 2 years old (Fig. 5.5(f)). However, seed production per tussock increased with time. The number of seedlings produced per tussock remained practically constant between 1969 and 1974 (Fig. 5.5(e)). The combination of these two trends suggests increasing seed mortality through the period.

Two other increases in mortality also occurred. Firstly, seedling mortality increased, slightly in percentage terms, but significantly in relation to recruitment. Secondly, mortality began to take a heavier toll among older individuals as well (Fig. 5.5(c)). These trends in mortality, survival and fecundity illustrate quite effectively how dynamic processes bring about local spatial and temporal (age-structured) distributions of individuals.

Perennial grasses growing in non-successional communities may also show regular recruitment from seed. Survivorship curves for several such populations have been constructed from annual or more irregular maps of seedlings and adult plant clumps in arid range grassland in the USA and Australia where an interest in the demography of these plants

Fig. 5.5 The population dynamics of *Corynephorus canescens*. (Data from Symonides 1979a)
(a) Map of tussocks and seedlings. (b) Seasonal fluctuations in total population density and annual density of adult plants. (c) Annual percentage mortality rate of seedling and non-seedling individuals. (d) Percentage of fruiting individuals in the non-seedling

population annually. (e) Annual fecundity of the adult population measured as seedlings/tussock and seeds/tussock. (f) Age structure of established plants.

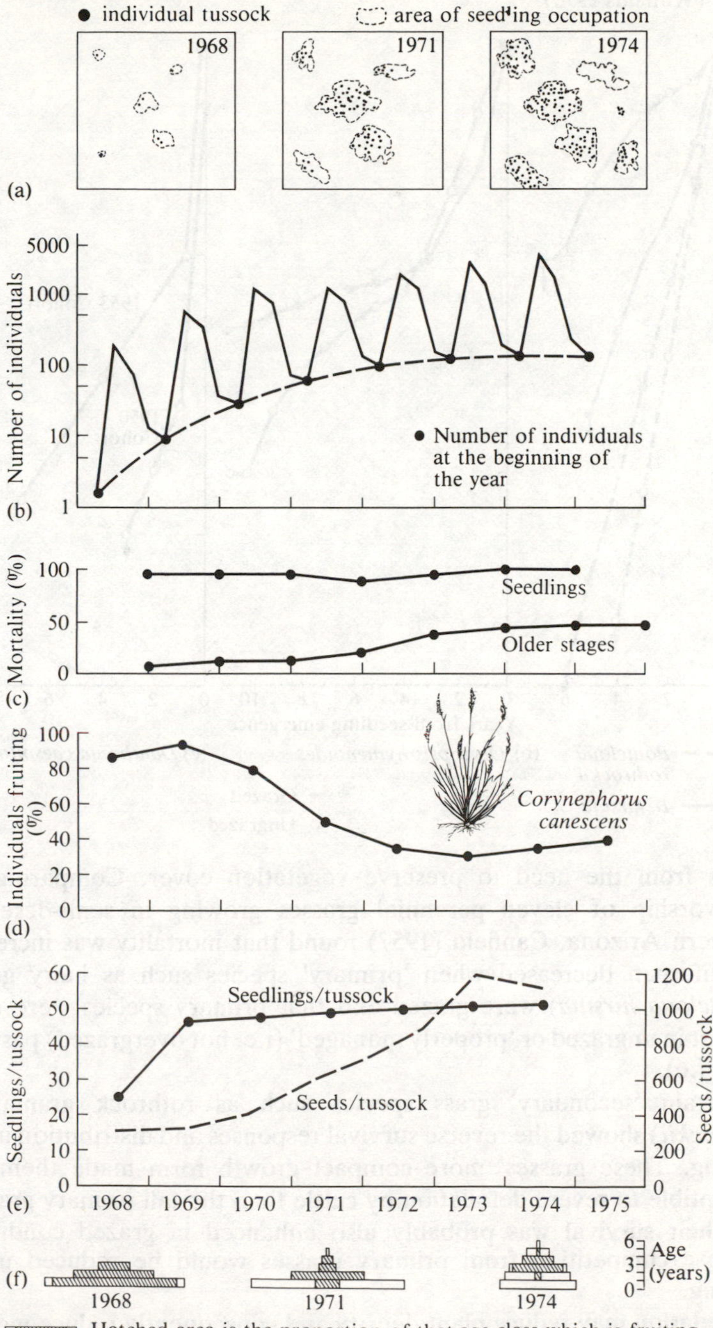

Fig. 5.6 The survivorship of some range grasses in grazed populations and ungrazed populations. Data: *Bouteloua* spp. (Canfield 1957); *Oryzopsis hymenoides* (West, Rea and Harniss 1979); *Danthonia caespitosa* (Williams 1970).

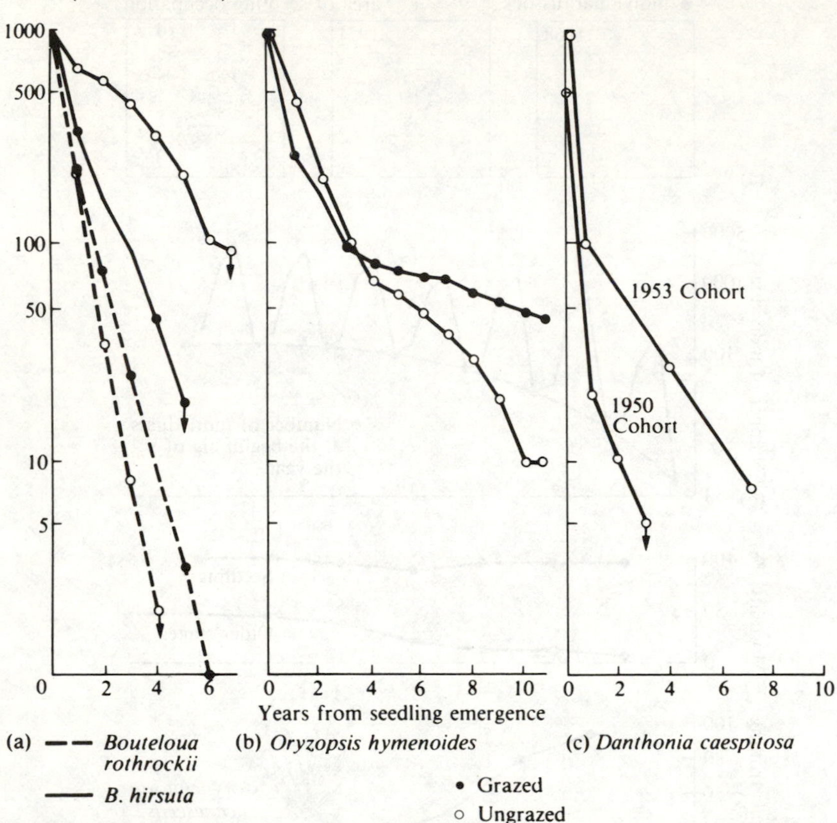

arises from the need to preserve vegetation cover. Comparing the survivorship of eleven perennial grasses growing in semi-desert in southern Arizona, Canfield (1957) found that mortality was increased and lifespan decreased when 'primary' species such as hairy grama (*Bouteloua hirsuta*) were grazed and that primary species were commonest in ungrazed or 'properly managed' (i.e. not overgrazed) pastures (Fig. 5.6).

Certain 'secondary' grass species such as rothrock grama (*B. rothrockii*) showed the reverse survival responses and distribution under grazing. These grasses' more compact growth form made them less susceptible to severe defoliation by cattle than the tall primary grasses, but their survival was probably also enhanced in grazed conditions because competition from primary grasses would be reduced under grazing.

Predation may reduce plant density and consequently reduce mortality caused by density stress in a number of situations. Seed predation by rodents was observed to have this effect in populations of annual grasses (Ch. 2, p. 27) and in experimental monocultures of shepherd's purse (*Capsella bursa-pastoris*) grazed by slugs (Dirzo and Harper 1980).

Autumn grazing has also been found to increase the survival of Indian ricegrass (*Oryzopsis hymenoides*) and several other grasses of sagebrush–grass communities in southeastern Idaho. West, Rea and Harniss (1979) explained this phenomenon on the grounds that these species are winter dormant so that grazers remove litter rather than green leaf when grazing at this time of year. This may stimulate later growth. Spring grazing of *Oryzopsis* would certainly decrease its survivorship.

In a meadow in North Carolina, Clay (1984) found that ramets of the grass *Danthonia spicata* infected with the parasitic fungus *Atkinsonella hypoxylon* had higher rates of growth and survival but lower fecundity than uninfected ramets and that infected genets were significantly larger than uninfected ones. Although trophically dependent upon grasses, the relationship of *A. hypoxylon* with these plants may be a mutualistic one, depending upon the relative contributions of fecundity and survival to fitness in *Danthonia* populations (Ch. 7). The balance between parasitism and mutualism may also be determined by grazing pressure since endophytic fungi in this group produce alkaloids which are toxic to herbivores and may deter them from consuming infected plants (Clay *et al.* 1985).

Other factors may also alter plant survivorship. Williams (1970) charted plants of an arid grassland species *Danthonia caespitosa* in Australia and found that cohorts arising in different years had differently shaped survivorship curves (Fig. 5.6c). These differences in survivorship are reminiscent of the long-term consequences which emergence order and initial rosette size were found to have on the survival of cohorts of *Viola blanda* and other plants (Ch. 2, p. 33). It is possible that the 1950 and 1953 cohorts of *Danthonia* were born into an environment sufficiently different to alter the initial growth size distribution of early survivors and hence also their later survivorship.

All these grass populations have fairly short half-lives of about 1–3 years and regular (often annual) recruitment from seed. This contrasts with the survival and recruitment of several perennial meadow and woodland herbs studied by Tamm (1956, 1972a, b) in Sweden.

In Tamm's initial study, populations consisted of plants of unknown age. They probably represented the accumulated survivors of many different years' seedling cohorts. Because of this uneven initial age structure it is inappropriate to describe the survival of such a population as survivorship, a term normally reserved for populations which fit the conventions of the life table. Nevertheless, the populations may be treated in a manner analogous to a proper cohort. The 'survivorship' curve for such a population is referred to as a *depletion curve* to distinguish it from true survivorship curves.

The half-lives of depletion curves, which are often exponential, are inevitably greater than they would be for the survivorship curves of true cohorts followed from the seedling stage because observations are begun with a set of 'proven' survivors.

Individual plants (probably mostly genets) of cowslip (*Primula veris*) mapped over a period of 29 years at one site had a depletion-curve half-life of 50 years, during which time no seedlings were able to establish and flower. The depletion curve for a population at another woodland site had a similar half-life for the first 14 years of the study but the population then abruptly nosedived with a half-life of 2.9 years when the tree canopy overhead became denser. The percentage of flowering rosettes also declined with age in this population although this trend was not observed in the more stable population. Depletion curves for some orchid populations studied by Tamm are shown in Fig. 5.7. Recruitment to the orchid populations in Tamm's plots was intermittent. This was a characteristic of many of the species he studied. Three other generaliza-

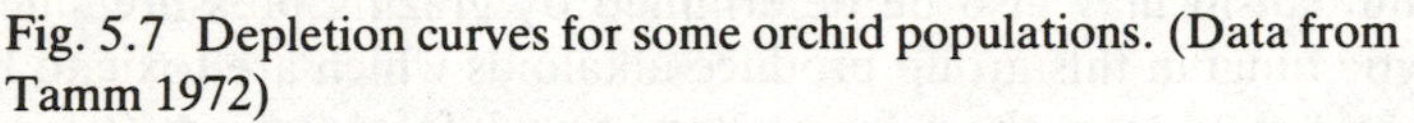

Fig. 5.7 Depletion curves for some orchid populations. (Data from Tamm 1972)

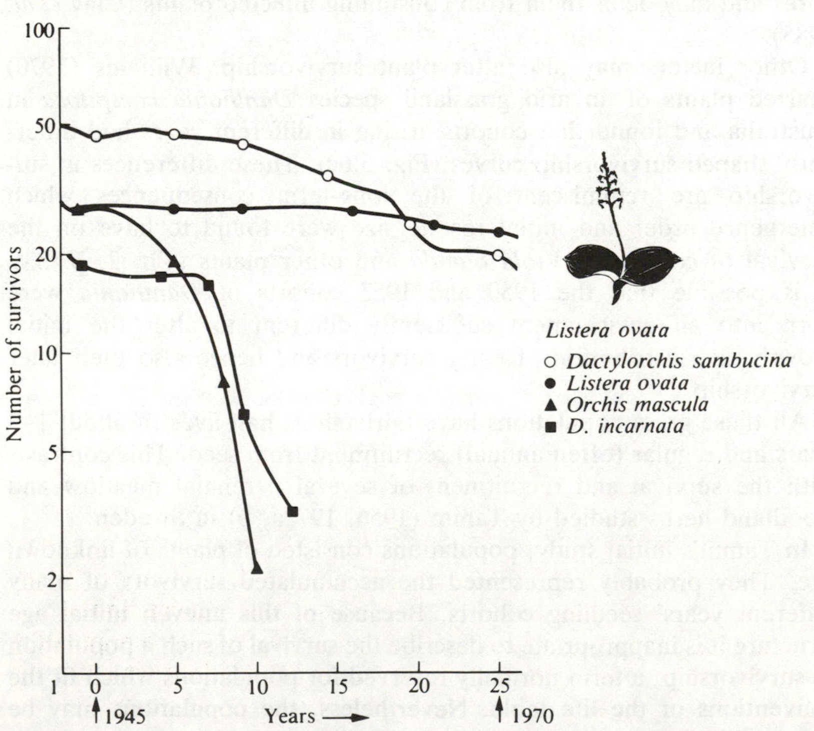

tions emerge from his work: 1. survival varies from species to species; 2. survival within a species varies from site to site (*P. veris*); 3. survival within a species at a particular site varies from time to time due to changing conditions.

The age of a number of perennial herbs can be determined from morphological indicators which mark annual growth increments. Thus although no dynamic life table may be available to indicate the frequency with which new recruits enter a population, stage structures or age structures may still provide some information of this sort retrospectively. The assumptions which allow static life tables to be built from age or stage structures are discussed further on page 97.

Bulbiferous species generally display a variety of life histories including regular recruitment. Of three woodland species of wild garlic studied by Kawano and Nagi (1975) in Japan one, *Allium victorialis*, displayed regular annual recruitment from seedlings, while two others possessed stage structures influenced predominantly by recruitment from vegetatively produced bulbils. In one of these species, *A. monanthum*, this produced a stage structure with many juvenile ramets. A stage or age structure for this population based upon genets would undoubtedly appear less youthful.

Stage structure may vary spatially within a population if recruitment and growth vary. The stage structure of *Erythronium japonicum*, a woodland lily studied by Kawano *et al.* (1982), showed a clear correlation with a gradient of light intensity (Fig. 5.8).

The flux of individuals through populations is of vital importance, particularly if rare species which appear to exist near the limits of extinction are to be properly managed and conserved. One of the first, and still one of the more thorough studies of population dynamics in herb populations is the one conducted by Sarukhan on buttercups, already referred to in Chapter 2.

A comparative study of three buttercups

The object of Sarukhán's study was to examine the flux of individuals through populations of *Ranunculus bulbosus*, *R. acris* and *R. repens* which occurred in an old pasture near Bangor in North Wales. The novelty of this simple objective may seem strange. However, just imagine being unable to answer the deceptively trivial question: 'How long does a buttercup live?' With the wisdom of hindsight the attentive reader would now reply: 'Do you mean a buttercup ramet or a genet?' The question is actually quite a complex one whose general and practical applications have already been emphasized.

Populations of the three buttercup species were monitored over a period of $2\frac{1}{2}$ years by mapping seedlings and rosettes occurring in permanent quadrats at intervals of a few weeks (Sarukhán and Harper, 1973).

Fig. 5.8 (a) The stage structure of *Erythronium japonicum* in square metre plots along a transect crossing the boundary between a deciduous wood (1–4 metres) and a conifer plantation (5–18 metres). Size classes are (0) seedlings, (1–12) non-flowering plants of successively larger size and (13) all flowering plants. (b) The gradient of light intensity along the transect, expressed as a percentage of full daylight. (From Kawano *et al.* 1982)

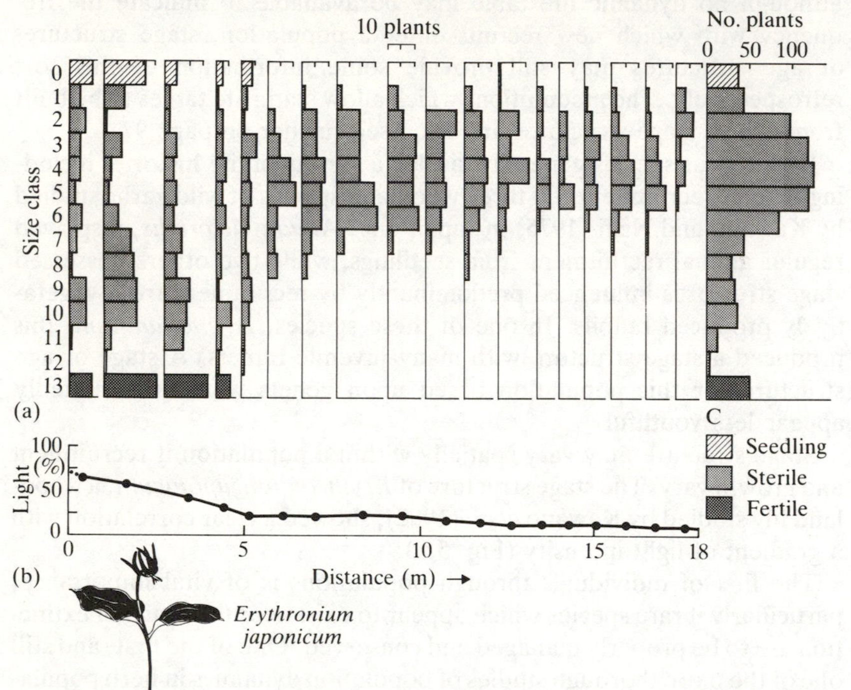

Rosettes of *R. repens* produce daughter rosettes (ramets) at the end of stolons, 20 cm or longer, which later wither, severing the connection between mother and daughter. *Ranunculus repens* also produces some seeds. *Ranunculus acris* produces seeds and also propagates vegetatively but produces fewer daughters than *R. repens*, and these grow on a stolon only a few millimetres from the parent. *Ranunculus bulbosus* produces seeds but has no vegetative propagation. Although it possesses a corm which is renewed annually, it is rare for a rosette to produce more than a single corm at a time.

The maps of these *Ranunculus* populations were used to follow the fate of over 9000 individual seedlings and rosettes and to draw up curves of the cumulative mortalities in populations from the beginning of the study in April 1969 (week 0), Fig. 5.9. All three species showed considerable turnover of rosette numbers. Further analysis of the population changes in *R. repens* showed that patterns of mortality were different for ramets and rosettes originating from seed (genets). The

Fig. 5.9 Population flux in *Ranunculus* species. (From Sarukhán and Harper 1973)

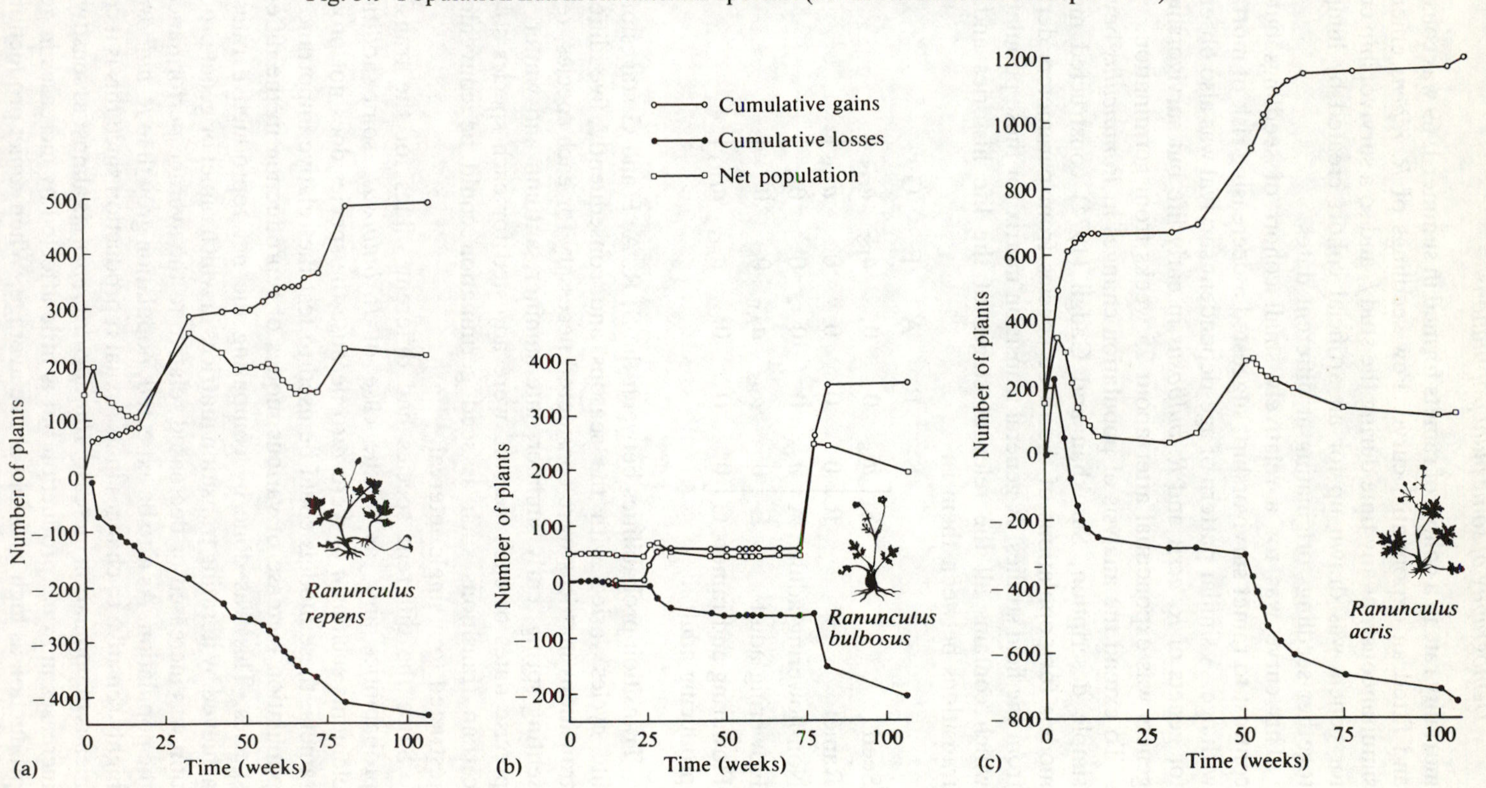

mortality rate in a cohort of ramets formed in summer 1969 was constant and fitted an exponential curve. Few seedlings of *R. repens* emerged simultaneously at any time during the study, and so a survivorship curve for genets was drawn up for an artificial cohort created by lumping together seedlings originating at different dates.

This curve was for a relatively small cohort of seedlings but, in contrast to ramet survivorship, suggested a decreasing risk of mortality with age. A similar pattern of age-dependent survival was also observed for genets of *R. acris* and *R. bulbosus* in early life but survivorship of genets was exponential after about 25 weeks from germination.

To extend the analysis of population changes in *Ranunculus* beyond simple description, Sarukhán and Gadgil (1974) constructed matrix models for populations of the three species from parameters derived from the field studies. A general transition matrix for these populations which contains all the relevant stages of the life histories and the transitions between them is:

		S	R	A	F	G
Seed	S	a_{SS}	0	0	a_{FS}	a_{GS}
Ramet	R	0	0	0	0	a_{GR}
Non-flowering adult	A	a_{SA}	0	0	0	0
Flowering adult	F	0	a_{RF}	a_{AF}	0	0
Flowering and ramet-producing adult	G	0	0	0	a_{FG}	a_{GG}

Transition probabilities between the S, R, A, F and G stages of the life cycles varied with the seasons and consequently five different transitions matrices were considered necessary for each species, representing spring, early summer, late summer, autumn and winter. The precise dates of each 'season' were adjusted for each species so that certain transitions such as seed germination could be conveniently restricted to a single interval.

Since the different species had different values for the transition probabilities and, as in the case of *R. bulbosus*, some additional elements such as a_{GR} were zero because the species does not produce ramets, these models could be used to test the relative importance for population increase of various modes of reproduction in the different species. This was done by comparing the net reproductive rates R_0 achieved by iterating transition matrices in which ramet or genet production was increased or decreased relative to the average performance of the population. As is to be expected, population growth in *R. bulbosus* is highly sensitive to changes in its sexual reproduction since this is its only means of population increase. *Ranunculus repens* is almost as sensitive to increases in *ramet* production but is little affected by increases in genet production as high as 3-fold above average. When *ramet* production by

R. repens is reduced below average R_0 is only slightly reduced because ramet mortality is density dependent. Population growth for *R. acris* is more strongly affected by increases in sexual reproduction than ramet production.

Sarukhán's censuses of the *R. repens* population were continued by Soane. Soane and Watkinson (1979) used 4 years' accumulated data to examine genet turnover and recruitment in this species in more detail than was possible within the more limited period of study. They found that genets (families of ramets) had approximately exponential survivorship. After 4 years the families surviving from the first year made a large contribution to the maintenance of the ramet population but a relatively small contribution to the number of different genets present. Though few in number, seedlings recruited since the first year made a significantly greater contribution to the genetic diversity of the population than the oldest, most prolifically ramifying genets.

A comparison of the demography of adjacent woodland and grassland populations of *Ranunculus repens* in North Wales was made by Lovett Doust (1981a) who found a higher density of ramets and a greater turnover in the woodland population. Using reciprocal transplants, she also demonstrated that there was some genetic differentiation between the populations and that, based upon several criteria, plants grew better on home ground than did transplants (Lovett Doust 1981b).

Genetic differentiation for life history characters affecting fitness was also found by van Groenendael (1985) in a comparison of populations of the plantain *Plantago lanceolata* growing in a dry grassland and a wet meadow. In this study too, natives had higher fitness than transplants from a different site. Using sensitivity analysis (Ch. 3) on the two populations, van Groenendael and Slim (in press) determined that recruitment from seed was more important to λ in the dry site than the wet one where conditions were less variable. Conversely adult survival was more important at the wet site.

Demographic differences between populations of a species at different sites are common, and may have important practical consequences, as the history of attempts to control ragwort with the cinnabar moth have shown.

Ragwort, the cinnabar moth and the role of insect herbivores in plant population dynamics

Senecio jacobaea ragwort (or tansy ragwort) is a short-lived perennial of sand dunes and heavily grazed pastures in Europe which has found its way into North America, Australia and New Zealand where it can be a serious pasture weed. This plant and its major herbivore the cinnabar moth caterpillar *Tyria jacobaeae* have been intensively studied in England and in Holland as well as in areas of their introduction. Collectively these studies demonstrate the importance of the spatial structure of plant populations for how they are regulated, as well as showing that population regulating mechanisms may be different in different places.

Ragwort plants establish from seed only in patches of bare ground. The seeds have a pappus that may disperse them on the wind for many metres, resulting in a uniform seed rain (Meijden *et al.* 1985). In their sand dune study site in Holland Meijden *et al.* (1985) found new populations recruited from both aerially dispersed seeds and the seed pool. Without herbivory, a ragwort plant develops a rosette, flowers within 2 or 3 years and then dies. This short life-cycle and the dependence upon vegetation gaps for re-establishment leads to a high turnover of local populations of *S. jacobaea* and wide fluctuations in numbers. The extinction rate varies from year to year but can be over 40 per cent.

Cinnabar moth caterpillars are patchily distributed and also fluctuate widely in numbers from year to year (Dempster 1982). They appear in groups on plants early in the year and consume flower heads and leaves, often stripping plants entirely. Defoliated plants regenerate from root buds which produce clusters of small rosettes. Since the caterpillars pupate in July or August and have only one generation per year, the new rosettes have time to grow before winter arrives, and may flower the following year. Defoliation followed by regeneration prolongs the life of populations attacked by *Tyria* in the Dutch dunes.

Despite the severe defoliation sometimes caused by *Tyria*, Dempster and Lakhani (1979) found that annual fluctuations in plant abundance at their British heathland site were closely correlated with rainfall and plant abundance in the year preceding each census. These two variables together accounted for 95 per cent of the variance in the abundance of *Senecio*. In some years new plants arose mainly from root buds and in others mostly from seed, but recruitment from both sources was similarly dependent upon rainfall. Though *Tyria* numbers did not affect plant abundance, they did influence the proportion of plants flowering in the year following defoliation. In Holland the situation appears to be somewhat different and, though attacked and non-attacked populations fluctuate in synchrony, testifying to the dominant influence of rainfall, *Tyria* numbers do have a 1-year-delayed effect on plant numbers there (Meijden *et al.* 1985).

Two kinds of *refuge* from mortality stabilize *Senecio* numbers in the Dutch dunes: 1. populations in areas shaded by trees have a lower extinction rate and also slower rates of increase than those in the open, providing a stable reserve of plants; 2. the egg-laying behaviour of female moths produces a highly aggregated distribution of eggs and caterpillars so that even some large populations of *Senecio* escape attack. Female selection of plants with an optimum nitrogen concentration in their leaves and avoidance of plants where predacious ants are present are probably both partly responsible for this pattern (Meijden *et al.* 1984; Meijden 1979).

It appears that an upper limit to population size in *Senecio jacobaea* is set by the availability of safe sites for establishment, a lower limit is set by refuges where local populations have a lower turnover, and rainfall determines density-independent fluctuations in recruitment. Herbivory plays, if any, only a minor role in population regulation of the plant, but at Dempster's study site the cinnabar moth is regulated by its food supply (Dempster 1982).

In North America the situation is similar, with an interesting exception. In most of the areas where *Tyria* has been introduced it has failed to control *Senecio jacobaea* because of the plant's ability to regenerate after attack (Myers 1980). Significantly, control has succeeded in Nova Scotia where summers are short and plants have insufficient time before winter to recover from herbivory (Harris *et al.* 1978). Situations of this kind where an insect herbivore is capable of regulating (near to extinction) a plant population appear to be the exception rather than the rule. The exceptions are often dramatic, such as the effect of the moth *Cactoblastis cactorum* upon *Opuntia* cacti in Australia, which turned large areas of cactus thicket into useful rangeland in the space of 5 years (Dodd 1940). Such projects involve insects introduced as biological control agents which have been carefully quarantined and imported without their parasitoids and diseases. However, even the most successful control agents do not cause total extinction of the plant. *Opuntia* spp. and *Cactoblastis* in Australia persist because the aggregated pattern of oviposition by the moth produces a refuge of plants that are not attacked, and leads to a low equilibrium density of plant and herbivore (Myers *et al.* 1981; Caughley and Lawton 1981).

Although, with the benefit of hindsight, it is possible to see why *Tyria* succeeded against *Senecio jacobaea* in Nova Scotia, there do not appear to be any reliable rules (beyond some obvious ones) for predicting which herbivore will control a particular plant in a particular situation (Julien *et al.* 1984; but see Ch. 6, p. 119). For example it was once thought that seed-attacking insects would be poor agents to use against weeds because such plants produce so many seeds that their populations are regulated by density-dependent seedling mortality. Hence any reduction in seed numbers should merely produce a compensatory reduction in mortality and have no net effect upon recruitment. The seed-eating weevil *Rhinocyllus conicus* which was introduced into Canada to control musk thistle *Carduus nutans* proves that idea wrong. Initially, the thistle colonizes bare soil in pastures. It is short-lived but produces permanent, impenetrable stands by self-replacement and by crowding out other plants where grazing by cattle is severe. Release of *R. conicus*, followed by controlled management of grazing to allow pasture species to recolonize, eliminates the thistle (Figure 5.10) (Harris 1984).

Not all successful biological control of weeds involves exotic herbivores. In a survey of control projects, Julien *et al.* (1984) found 31 which

Fig. 5.10 Population dynamics of *Carduus nutans* and a seed-eating weevil *Rhinocyllus conicus* following its introduction as an agent of biological control in 1968 and 1969 at Aylesbury, Saskatchewan, Canada. (Data from Harris 1984)

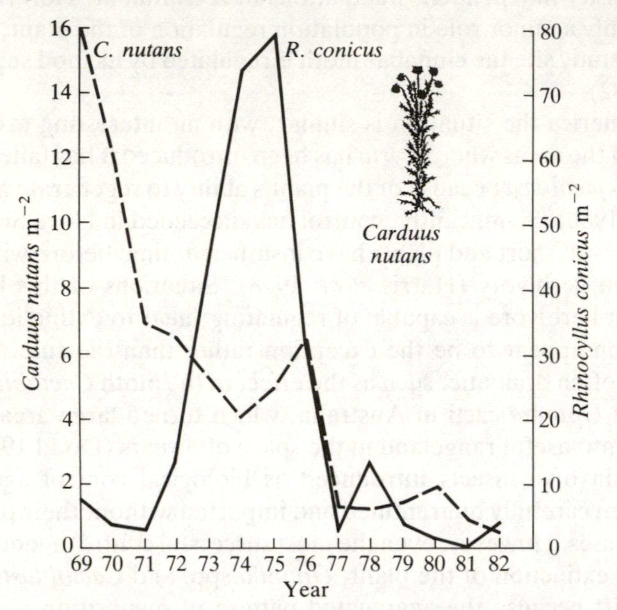

had used a native organism, of which 18 (58%) achieved some success in control. This is actually a higher success rate than for non-native agents (47%), perhaps because larger populations of native organisms are generally used.

We know that *Cactoblastis* plays an important role in the population dynamics of *Opuntia* in Australia because the introduction of the moth there constituted an experimental test of this on a giant scale. We cannot know whether insect herbivory plays a role in the dynamics of native plant species unless insects are experimentally removed. The effect of insect herbivory upon seed set and plant survival in the shrubs *Haplopappus squarrosus* (Louda 1982b) and *H. venetus* was investigated in this way by Louda (1982a,b, 1983) at a series of sites along a gradient from the California coast, inland.

Applications of insecticide significantly increased the production of viable seed by both species at all sites. Juvenile mortality of both species was density independent and so seedling numbers increased in proportion to seed production. In *H. squarrosus* juvenile mortality was fairly low and did not vary along the gradient, so that seed set, and the limit set upon it by predation, was the principal factor influencing the relative numbers of plants recruited along the gradient. In *H. venetus*, juvenile mortality was much greater and hence the increase in seedling density in insecticide-treated plots was insufficient to alter recruitment which remained low. In this species seed predation apparently did not influence differences in plant abundance which occurred along the gradient.

Grazing by vertebrates is frequently a dominant influence upon the population structure of shrubs, particularly in arid regions where they are important constituents of otherwise sparse vegetation. We look at these plants in the following section.

Shrubs and the analysis of age structure

Many shrubs such as creosote bush (*Larrea tridentata*) which is an important species in the southwestern deserts of North America spread clonally. This aspect of shrub demography is dealt with in Chapter 6. Grazing by rabbits and sheep was responsible for major differences in the survivorship of the desert shrub *Cassia nemophila* between enclosed and unenclosed areas of the Koonamore Reserve in South Australia, though the biggest losses in both populations occurred in age classes less than 5 years old (Silander 1983). Survivorship of the shrubs *Acacia burkitii*, at the same reserve, and *Artemesia tripartita* at a similar site in Idaho are shown in Fig. 5.11.

Broom (*Sarothamnus scoparius*) is a European shrub of mesic grassland and heathland habitats. Comparing its population age structure at two sites in France, Rousseau and Loiseau (1982) found that grazing by pasture animals had a major impact upon recruitment and adult survivorship. The effect of insect herbivores on a cultivated population of *S. scoparius* has been studied in some detail by Waloff and Richards (1977) in Berkshire, England. They transplanted 240 seedlings into two plots in which their mortality and seed production were followed for 11 years. The plants in one plot received a regular insecticide treatment to the foliage and surrounding soil which substantially reduced the populations of phytophagous insects on them. Thirty species of insect attacked shrubs in the unsprayed plot, including aphids on leaves and seed pods, lepidopteran stem-miners, seed weevils, agromyzid leaf-miners, beetles feeding on the flowers and a weevil whose larvae feed on root nodules and whose adults feed on broom leaves. Under pressure from this onslaught shrubs in the unsprayed plot suffered twice the mortality of those treated with insecticide (Fig. 5.11), and in 10 years produced only 25 per cent of the seeds per bush produced by sprayed shrubs.

Parker (1985) found that a beetle attacking the roots of the composite shrub *Gutierrezia microcephala* in New Mexico only infested large individuals. Clearly, the age structures of shrub (and other) populations may reflect the impact of invertebrate as well as vertebrate herbivory. In temperate woody plants which have annual rings in the trunk, various refinements have been devised for obtaining the most accurate estimate of an individual's age (Roughton 1962). The end-product of all methods is a population age structure, equivalent to the column matrix in a matrix

Fig. 5.11 Survivorship curves for three shrub populations. Data: *Sarothamnus scoparius* (Waloff and Richards 1977); *Acacia burkittii* (Crisp and Lange 1976); *Artemisia tripartita* (West, Rea and Harniss 1979).

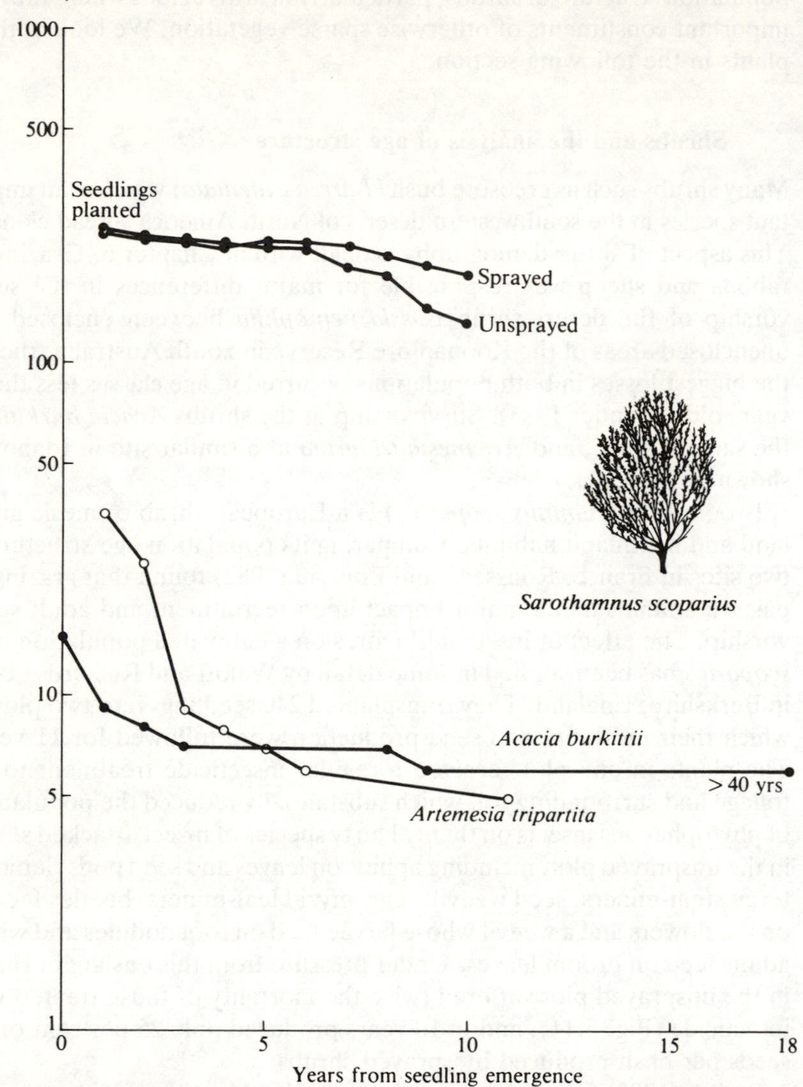

model (Ch. 3). An example of such an age structure for *Acacia burkittii* is shown in Fig. 5.12. An age structure is like a single frame of a movie film that can give us a clue about what came before and what might follow if we know how to interpret it. Any given age structure can be reconstructed if we assume the correct age-specific mortality and natality that operated during earlier frames of the movie. Taking a simple, hypothetical example, the structure in Fig. 5.13(c) can be explained with the following

Fig. 5.12 Age structures for two contrasting populations of *Acacia burkittii*. (Crisp and Lange 1976)

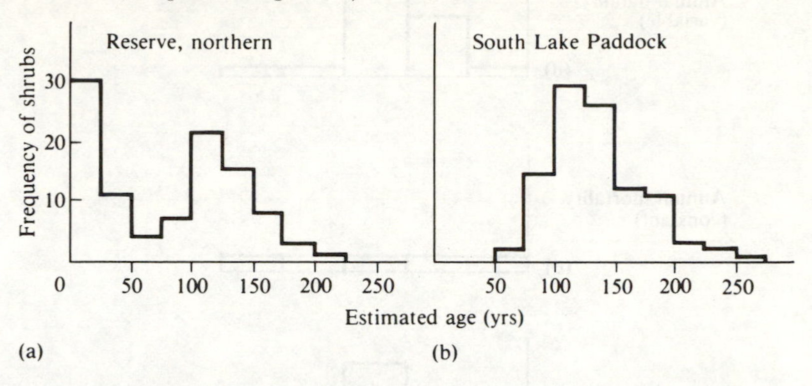

assumptions: 1. natality was constant each year (Fig. 5.13(a)); 2. mortality varied each year (Fig. 5.13(b)); 3. mortality acted on first-year seedlings only. Equally, the age structure of our hypothetical population could have been produced by variable annual natality (Fig. 5.13(d)) and a constant, low annual mortality (Fig. 5.13(e)).

There are, of course, combinations of varying natality and varying mortality between the extremes we have illustrated which could also produce the observed age distribution.

These models assume that the number of individuals now 5 years old is determined by the number born 5 years ago minus the number of those that died in that year. It makes no allowance for those in the

Fig. 5.13 A hypothetical age structure (c and f) is the product of annual natality minus annual mortality. See text for further explanation.

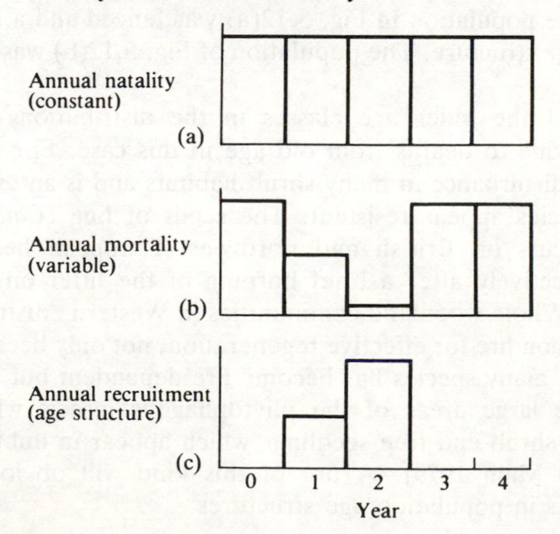

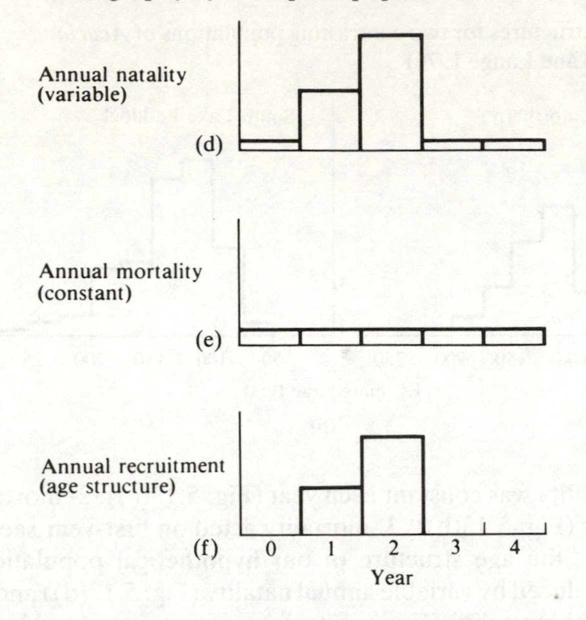

original cohort that died in their second or third year. This may be a more reasonable simplification than at first appears because some, probably many, plants do have massive juvenile mortality. How reasonable is the other assumption about seedling mortality, that it has been very variable in the history of the population? The answer to this question must, of course, depend on the habitat but there are many cases of peaks in population age structures coinciding with historically dated episodes such as a reduction in grazing or the incidence of fire. The dearth of *Acacia* plants younger than 100 years in Fig. 5.12(b) (p. 99) coincides with the introduction of sheep to the area. About 50 years before these shrubs were aged the population in Fig. 5.12(a) was fenced and a new peak is seen in its age structure. The population of Fig. 5.12(b) was left unfenced.

The truncation of the older age classes in the distributions are presumably simply due to deaths from old age in this case. Fire is a common source of disturbance in many shrub habitats and is an event to which many species appear resistant. The seeds of ling (*Calluna vulgaris*) which occurs on British and northwest European heaths germinate most effectively after a brief burning of the litter on the surface of the soil. Whole woodland communities in Western Australia appear to depend upon fire for effective regeneration, not only because seed germination in many species has become fire dependent but also because fires purge large areas of the phytophagous insects which consume any herb, shrub and tree seedlings which appear in unburnt areas (Whelan and Main 1979). A fire of this kind will obviously produce sharp peaks in population age structures.

Fig. 5.14 An hypothetical age structure resulting from constant annual natality and 50 per cent mortality per year.

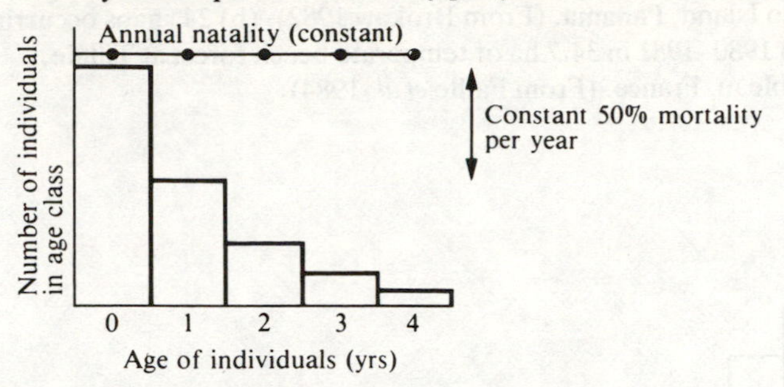

Variations in mortality and natality may produce peaks in age structures by truncating younger age classes. What would the age structure of a population with constant annual natality and some constant annual value of mortality (say 50%) look like?

Assuming a constant yearly input of seedlings to the 0 age class, a constant percentage mortality per year would produce an age structure with an exponential profile (Fig. 5.14). A histogram of a cohort of seedlings with exponential survivorship drawn from a dynamic life table would look just like Fig. 5.14. In fact the assumption of constant annual natality enables us to treat a static life table based on age structure as though it were a dynamic life table. The age structure of some high-altitude *Abies* populations (e.g. Fig. 4.13, p. 69) can be interpreted in this way as we will see in the next section.

Forest trees

The key event in the demography of trees in both temperate and tropical forests is the appearance of gaps in the canopy and the minor and major disturbances which create them. Most gaps are small (Fig. 5.15), caused by the fall of individual branches, and these are filled rapidly by the crowns of bordering trees. Larger gaps are filled by germination from the seed pool and by recruitment from tree seedlings (oskars, Ch. 2) already present on the forest floor. In small and intermediate sized gaps in beech–maple forest in New York state, Canham (1985) found that sugar maples (*Acer saccharum*) experienced several episodes of suppression and release before reaching the canopy. Oskars of a tropical tree in the genus *Aglaia* growing in Malaysian dipterocarp rainforest also needed several openings of the canopy to reach maturity (Becker and Wong 1985).

The size of a gap, and when and how it is created are important factors determining the course of subsequent colonization. Trees depending upon oskars are termed 'primary' species. Larger gaps admit 'pioneer' tree species such as *Prunus pensylvanica* in eastern North American

Fig. 5.15 The size-distribution (m^2 surface area) of forest gaps: (a) 66 gaps occurring between 1975–1980 in 13.4 ha of tropical forest on Barro Colorado Island, Panama. (From Brokaw 1982). (b) 243 gaps occurring between 1980–1981 in 34.7 ha of temperate beech forest at Tillaie, Fontainbleau, France. (From Faille *et al.* 1984)

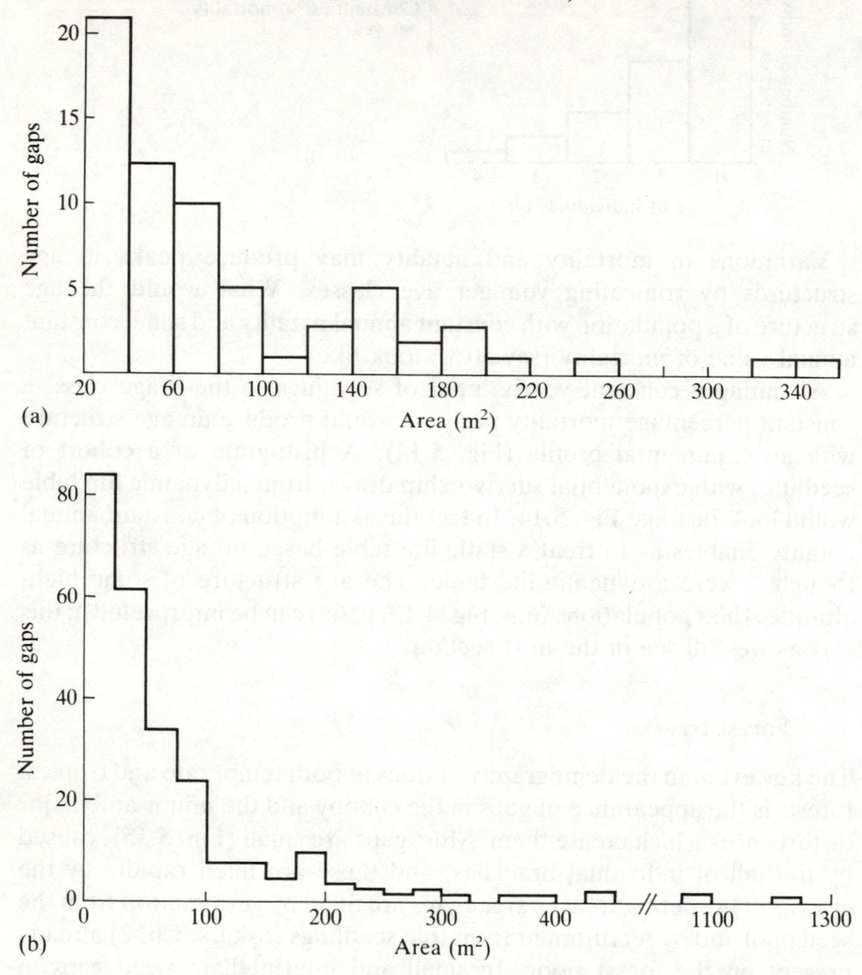

forests (Marks 1974) or *Cecropia* spp. in the neotropics (Vazquez-Yanes and Smith 1982). These species lack oskars but have a pool of buried seeds that may be unearthed from some depth by the upheaval of roots and soil when a large tree falls (Putz 1983). In gaps studied by Brokaw (1985) at Barro Colorado Island (BCI) in Panama, rapidly growing pioneers dominated the early phase of colonization in large gaps but were outlived by primary species (Fig. 5.16(a) and (b)). Larger, longer-lasting gaps allow a longer period of recruitment and were colonized by a greater diversity of species than small gaps in several eastern North American forests studied by Runkle (1982).

Fig. 5.16 Changes in tree density with time: (a) ○, primary species and ●, pioneer species in a small (20 m^2) gap and (b) in a large gap (532 m^2) in tropical forest at BCI. (From Brokaw 1985). (c) Density of all species in gaps of different age in evergreen oak forest in Japan. (From Naka and Yoneda 1984)

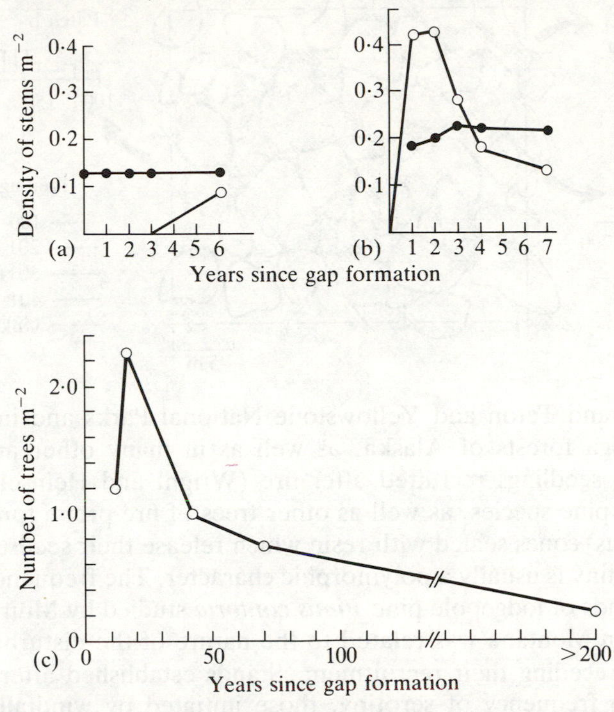

The high densities of trees recruited in gaps cause density-dependent mortality and a reduction in numbers with time (Fig. 5.16(c)). Mortality of saplings caused by disease is often higher near parents or other adults of the same species, though this effect is weaker while a canopy gap remains open, admitting light (Ch. 9, p. 190). In natural forests the dependence of both pioneers and primary species upon gaps creates a mosaic pattern of species composition and of age structure. The latter is seen in Fig. 5.17 which shows the age-class and spatial distribution of trees in a *Tsuga–Abies–Picea* subalpine forest in Japan. Although trees of similar age occur together in patches, each patch is not an even-aged cohort because these primary species began life as oskars at different times before the gap (or gaps) which finally allowed them to mature.

Wildfire is a common source of natural disturbance in many forests. Its role is best documented in North America where conifer species in particular appear to depend upon it for regeneration. A natural 'fire rotation' of about 100 years is thought to have prevailed in pine forests in Minnesota before settlement times (Heinselman 1973; Swain 1973). Mature trees of forests in the Sierra Nevada, California, the northern Rocky

Fig. 5.17 The distribution of trees in a 30 × 30 m plot in a subalpine *Tsuga–Abies–Picea* forest in Honshu, Japan. The crown of each tree is outlined. Patches of trees of similar age are marked and their age-structure is shown in histograms. (From Kanzaki 1984)

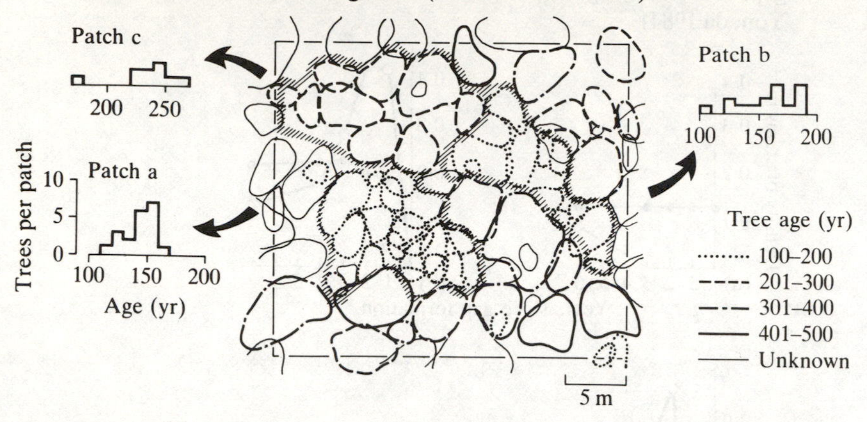

mountains, Grand Teton and Yellowstone National Parks and in the boreal and taiga forests of Alaska, as well as in many other areas, originate from seedlings recruited after fire (Wright and Heinselman 1973). Several pine species, as well as other trees of fire-prone forests, have 'serotinous' cones sealed with resin which release their seeds only after fire. Serotiny is usually a polymorphic character. The frequency of serotiny in stands of lodgepole pine *Pinus contorta* studied by Muir and Lotan (1985) in Montana was related to the nature of the disturbance immediately preceding their recruitment. Stands established after fire showed a high frequency of serotiny, those initiated by windfalls or disease showed a low frequency. This is an example of disruptive natural selection powerful enough to alter gene frequencies in a single generation.

Clearly, the demography of forest trees and the fate of individuals has to be viewed in the context of natural forest disturbance. An unusual situation, characteristic of high altitude *Abies* forests, allowed Kohyama and Fujita (1981) to reconstruct a survivorship curve for *Abies* spp. from the age structure of stands of known age (Fig. 5.18(a)). Bands of mortality advance like waves through these forests, killing trees along a broad front. Even-aged cohorts of *Abies* are recruited in the gaps which occur in the wake of these waves. Because recruitment is effectively invariable through time and trees occur in even-aged cohorts, an age structure for such a forest (e.g. the one for *Abies balsamea* shown in Fig. 4.13, p. 69) can be interpreted like a dynamic life table (p. 101 this chapter).

The abrupt increase in mortality at about 80–90 years of age (Fig. 5.18(a)) when the mortality wave arrives is peculiar to high altitude wave-regenerating *Abies* populations but in other respects this survivorship curve is fairly typical of trees. Heavy seed and seedling mortality

Fig. 5.18 Survivorship curves for: (a) all trees in a mixed stand of *Abies veitchii* and *A. mariesii*, Mt. Shimagare, Japan (from Kohyama and Fujita 1981); and (b) for cohorts of seeds and seedlings of *Acer saccharum* (Curtis 1957 and Hett 1971) and *Pinus sylvestris* (Guittet and Laberche 1974), and seedlings of *Shorea parviflora*. (Wyatt Smith 1958)

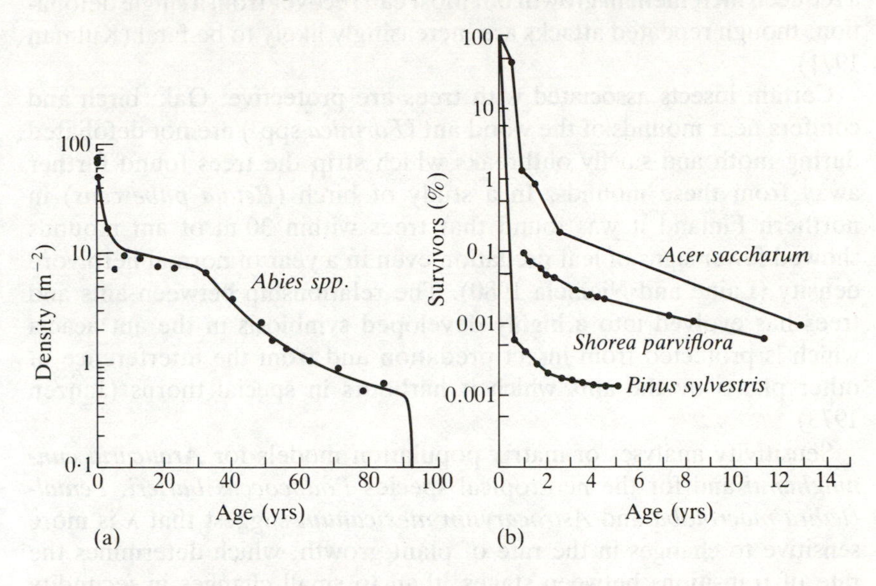

and an adult mortality rate that declines with age is usual (Fig. 5.18(b)). Age structures implying this pattern have been identified in many forest tree populations including striped maple *Acer pensylvanicum* (Hibbs 1979), balsam fir *Abies balsamea* and eastern hemlock *Tsuga canadensis* (Hett and Loucks 1976) in North American forests; Chile pines *Araucaria* spp. in New Guinea (Gray 1975) and many neotropical trees.

The importance of density as a cause of mortality in crowded populations of seedlings has already been mentioned. The incidence of diseases, particularly fungal pathogens, frequently increases with the density of a plant population because pathogens are able to spread more quickly and easily. However, this may not apply where the pathogen has an obligate alternate host. *Cronartium fusiforme* fusiform rust of pines is more common in sparse pine stands than dense ones because the understorey oaks which are the obligate alternate host of the rust are more common in these than in dense stands. Pines in dense stands may also be less susceptible to infection because they shed lateral branches, which are the sites of infection, earlier than open-grown trees (Burdon and Chilvers 1982).

The diseases and insects which attack trees are often age specific (Kulman 1971; Gray 1972). Seedlings are prone to fungal 'damping-off' disease for instance while defoliation of tamarack (*Larix laricina*) by the larch bud moth may affect mature trees in a reproductive state more severely than juveniles (Niemela, Tuomi and Haukioja 1980). Other diseases attack trees already weakened by storms or insect attack. Moth caterpillars are often responsible for the severe defoliation of trees. Insects such as the winter moth, gypsy moth and the spruce budworm experience population explosions in certain years. Defoliated trees show a reduced increment in growth but most can recover from a single defoliation, though repeated attacks are increasingly likely to be fatal (Kulman 1971).

Certain insects associated with trees are protective. Oak, birch and conifers near mounds of the wood ant (*Formica* spp.) are not defoliated during moth and sawfly outbreaks which strip the trees found further away from these mounds. In a study of birch (*Betula pubescens*) in northern Finland it was found that trees within 30 m of ant mounds showed fewer signs of leaf predation even in a year of normal herbivore density (Laine and Niemela 1980). The relationship between ants and trees has evolved into a highly developed symbiosis in the ant acacia which is protected from insect predation and from the interference of other plants by the ants which it harbours in special thorns (Janzen 1973).

Sensitivity analyses of matrix population models for *Araucaria cunninghamii* and for the neotropical species *Podococcus barteri*, *Pentalclethra macroloba* and *Astrocaryum mexicanum* suggest that λ is more sensitive to changes in the rate of plant growth, which determines the rate of transitions between stages, than to small changes in fecundity or survival in these trees (Pinero *et al.* 1984). In *A. mexicanum* an individual's growth rate and fecundity are related to the light environment in the forest understorey where it grows rather than to its age (Pinero and Sarukhan 1982), indicating that even in a 'shade tolerant' species its population dynamics will depend upon the frequency and size of gaps in the canopy.

We have already, unavoidably, dealt with the demography of some clonal plants such as *Ranunculus repens* but these deserve further attention in the next chapter.

Summary

Vertical or *static life tables* are compiled from age structures, horizontal or *dynamic life tables* from following the survivorship of a cohort. The shape of a survivorship curve is influenced by the life history characteristics of a species and by local environmental conditions, including *density*, *disease* and *herbivory*. The causes of mortality may vary with the age of plant and with season. In a number of herb species, the survival and reproduction of plants in *transplant experiments* has shown them to be adapted to local conditions at the home site.

Although insect herbivory may significantly affect the fitness of individual plants, it need not alter population size. Experiments are needed to elucidate the effect of native insect herbivores upon plant populations. The *spatial distribution* of plants may provide a *refuge* from herbivores and stabilize population numbers.

A reserve of *dormant seeds* and/or regular environmental *disturbance* are necessary for the persistence of populations of short-lived plants. Recruitment from seed is more intermittent among many herbaceous perennials and trees but environmental disturbance and *canopy gaps* are still important to them.

6
Clonal plants

Plants are modular in construction and thus have an open-ended pattern of growth. A tree adds modules to its branches and so extends vertically, a clonal plant adds modules at its base and extends horizontally. A few clonal shrubs and trees do both. The difference between these growth habits is mostly one of scale.

Clonal growth may be defined as the horizontal extension of a plant by the addition of ramets which develop their own roots. It has been estimated that more than two-thirds of the commonest perennials in the British flora show pronounced clonal growth (Salisbury 1942). Even greater proportions of woodland and aquatic herbs show this habit. The potential efficiency of vegetative propagation in aquatic herbs has been demonstrated by the spread of Canadian pondweed (*Elodea canadensis*) after its introduction into Britain at Market Harborough in 1845. From this introduction the species rapidly colonized most of Britain's waterways entirely by the proliferation of ramets which severed, spread and further multiplied. The species is almost exclusively female in Britain and sexual reproduction is very rare or does not occur.

Many plants have structures such as the stolons of strawberries or the rhizomes of bracken which carry ramets into the space around the mother plant. The fact that these vegetative daughters may lead an independent existence has lead to the widespread description of this process as *vegetative reproduction* (e.g. Abrahamson 1980). Since the daughters of 'vegetative reproduction' are not only identical to their mother but are often actually physically and physiologically connected to her, the use of the term *reproduction* seems inappropriate in this context and we will use the term clonal growth instead.

There are examples of every degree of separation between mother and daughter from the tillering of grasses such as *Festuca rubra* in which tightly packed clumps of ramets are formed, to the spreading habit of white clover (*Trifolium repens*) which invades the surrounding turf with stolons which widely disseminate the shoots that arise from them (Table 6.1). In the latter species the mother shoot has a short life and the combined effect of these maternal deaths, the birth of new shoots and the stoloniferous habit is that white clover wanders about in a grass sward or advances centrifugally forming rings of clover in the turf.

Table 6.1 Characteristics of some clonal perennial herbs

Habitat and species	Longevity of ramets (yr)	Duration of connections (yr)	Annual growth of rhizome etc. (m)	Translocation between ramets	Sources
Forest understorey					
Aralia nudicaulis	26	Permanent	0.25–0.80	—	1, 1, 2
Aster acuminatus	1	< = 2	0.01–0.25	No	3, 4, 2, 4
Clintonia borealis	1	10–12	0.06–0.10	Yes	5, 5, 2, 4
Fragaria virginiana	2–6*	< 1	< = 1	Yes	6, 6, 6, 7
Medeola virginiana	1	< 1	0.02–0.08	—	8, 8, 2
Mercurialis perennis	1	5–6	0.10–0.40	—	9, 9, 10
Narcissus pseudonarcissus	> = 4	2–3	0.01	—	11, 11, 11
Trientalis borealis	1	1	0.5–< 1	—	12, 12, 12
Viola blanda	4	4	0.6–0.35	Yes	13, 13, 13, 14
Open habitats					
Elymus repens	1	> 1	0.5–2	Yes	15, 15, 15, 16
Carex arenaria	< = 2	> = 7	3.6	Yes	17, 17, 17, 18
Phleum pratense	< = 1	> 1	—	No	20, 20, –, 21
Ranunculus repens	1.2–2.1	< 1	0.15–0.50	Yes	22, 22, 22, 23
Solidago canadensis	1	> 3	0.01–0.15	Yes	24, 25, 24, 26
Trifolium repens	—	2	0.4	Yes	–, 27, 27, 28

–, No data
* Half-life of a cohort of ramets

Sources: [1] Edwards (1984), [2] Sobey and Barkhouse (1977), [3] Pitelka *et al.* (1980), [4] Ashmun *et al.* (1982), [5] Pitelka and Ashmun (1986), [6] Angevine (1983), [7] Jurik (1985), [8] Bell (1974); Cook (1986), [9] Falinska (1982), [10] Mukerjii (1936), [11] Barkham (1980); Barkham and Hance (1982), [12] Anderson and Loucks (1973), [13] Cook (1983), [14] Newell (1982), [15] Palmer (1958), [16] Rogan and Smith (1974), [17] Noble *et al.* (1979), [18] Tietema (1980); Noble and Marshall (1983), [19] Bishop *et al.* (1978), [20] Langer *et al.* (1964), [21] Williams (1964), [22] Sarukhan and Harper (1973), [23] Ginzo and Lovell (1973), [24] Bradbury (1981), [25] Bradbury and Hofstra (1977), [26] Hartnett and **Bazzaz** (1983), [27] Erith (1924), [28] Ryle *et al.* (1981).

The spatial organization of clones

The simplest pattern of clonal growth is where a genet consisting of a single rhizome advances by increments in a straight line. A plant with this one-track approach to life, sand sedge (*Carex arenaria*), is found in sand dunes where its predominantly linear structure is put to efficient use as it invades areas of uncolonized sand (Fig. 6.1). The rhizome grows by the repeated production of a module with four nodes (leaf scars) ending in two underground buds. One of these produces a new underground module and the other may remain dormant or produce an aerial shoot with its own dormant bud. This may occasionally produce a rhizome branch. Shoots themselves go through a series of growth stages. Some recruitment and mortality occurs in all months of the year but both are mostly confined to the summer months in British populations.

Fig. 6.1 The invasion of a plot by *Carex arenaria*. (○) New shoots; (●) old shoots. The relative proportion of old and new shoots is shown next to each annual chart of the study plot. (From Symonides 1979b)

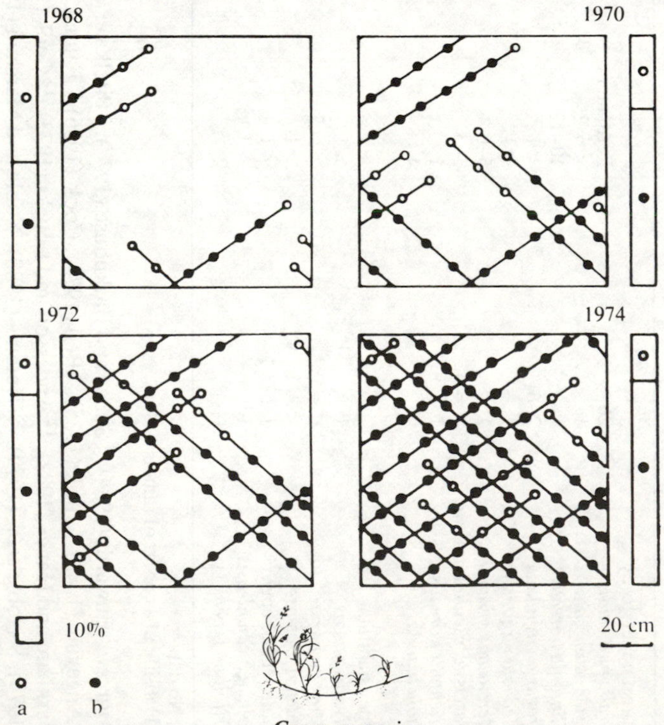

As an area is colonized the shoot stage structure shifts from juvenile to senescent.

Rhizome buds may remain dormant long after the shoots with which they were produced have died and they may accumulate in numbers up to 400–500 m^{-2} (Noble, Bell and Harper 1979). This pool of dormant buds, analogous to the pool of dormant seeds found in other species, may form the basis for recruitment to the shoot population where new sand accumulates over an area. In a Welsh population of *C. arenaria* studied by Noble, Bell and Harper (1979), fertilizer was applied to some sites containing buried rhizomes with dormant buds. The flux of aerial shoots was compared in these and in untreated sites. Both shoot birth- and death-rates rose in the fertilized populations, but the net shoot population also increased substantially. The age structure of fertilized populations was rejuvenated as rapidly senescing older shoots were replaced by new ones.

The linear rhizome structure of *C. arenaria* seems particularly well suited to the marginal invasion of mobile sand. It has been suggested by Bell and Tomlinson (1980) that other specific rhizome arrangements also permit plants of other habitats to exploit the space available to them in the most efficient way. There are examples of plants with regular branching angles in many unrelated families (Bell and Tomlinson 1980), and among them one of the commonest angles appears to be 60° (e.g. *Solidago canadensis*, Smith and Palmer 1976; *Alpinia speciosa*, Bell 1979). This particular angle produces hexagonal rhizome arrangements which are among the most efficient geometric shapes for close-packing a two-dimensional space. It seems likely that this may be an evolutionary solution to the problem many clonal plants face which is to exploit available space to the full without crowding themselves. On the other hand it has been pointed out (Mortimer 1984; Cook 1986) that excavated rhizome systems are rarely as orderly as one would expect if these plants really behaved as the 'rules' constructed for them by ecologists demand!

Clonal growth can produce linear arrangements, networks or clumped distributions of ramets. It is not at all clear why there should be such a range of clonal behaviour, even within forest herbs where, for example, *Medeola virginiana* regularly fragments but many other species do not (Table 6.1) (Cook 1983; Hutchings and Bradbury 1986). In most clonal species in which the rhizome, stolon or other connection remains intact, the genet behaves as a physiologically integrated unit. All clonal species, however they are structurally organized, can perhaps be regarded as 'foraging' for resources after their own fashion (Harper 1986). In *Viola blanda* (Cook 1983), *Aralia nudicaulis* (Edwards 1984) and some other species (Kershaw 1962) ramets form clumps, though excavation has revealed that these are frequently formed by the convergence of rhizomes or stolons from different genets. Cook (1986) suggests that stolons of

Viola blanda may respond to local pockets of nutrients, initiating new ramets in the vicinity of rotting wood. Rhizomes of couch grass *Elymus repens* may behave the same way in response to patches of soil rich in nitrogen (Mortimer 1984). In *Aralia nudicaulis* and other forest floor species, clonal growth may aid plants in locating light-gaps, as the understorey runners of the tropical vine *Ipomoea phillomega* (Ch. 1, Fig. 1.3) evidently do.

The rhizomatous perennial ragweed *Ambrosia psilostachya* occurs in saline and non-saline habitats in the Great Plains of western USA (Salzman and Parker 1985). Genotypes of this plant vary in their ability to survive and grow in saline soil. In experiments where plants were grown in the centre of pots containing a soil salinity gradient, all genotypes produced most of their rhizomes in the less saline parts of the pots, but this 'preference' was much more marked in less salt-tolerant genotypes which, in effect, avoided saline habitat by selective rhizome initiation. In the field, the rhizomes of experimental transplants into saline soil grew significantly longer than those in non-saline soil, suggesting that this flexibility in clonal growth helps plants 'escape' saline microhabitats and locate less saline ones (Salzman 1985).

The demography of ramets

Populations of ramets have birth, death and reproductive rates just as genets do. However, it is important to distinguish between these processes in genets and ramets because the behaviour of a set of units which are physiologically integrated with each other to some degree is likely to be different from the behaviour of unconnected genets. In an experiment, Hartnett and Bazzaz (1985c) found that interconnected ramets of goldenrod (*Solidago canadensis*) were less affected by intraspecific competition at high density than were individual shoots grown from seed. This may have been because the connected shoots possessed a rhizome which contained stored assimilates unavailable to the single shoots. Storage allows a clone to even out the bad times with the good.

The ability to translocate assimilates from ramet to ramet allows a clone to even out spatial heterogeneity too. In a different experiment, Hartnett and Bazzaz (1985b) found that *S. canadensis* ramets growing in backgrounds of other species produced fewer leaves with some neighbour species than others. When interconnected ramets were grown so that each ramet had a different neighbour species, ramets did not show a specific response to neighbours but appeared to average out the effect of different neighbours between them. However, to be sure that physiological integration is responsible for this result, the experiment would have to be repeated with unconnected ramets because an averaging effect could also result from the direct influence of a variety of neighbours upon the *aerial* part of each *individual* ramet. In a similar experiment with

Ambrosia psilostachya, only the below-ground environment of ramets was altered, thus avoiding the confounding of aerial effects with subterranean ones. Salzman and Parker (1985) grew *A. psilostachya* in pots with two compartments. For a two-environment treatment, one compartment contained saline soil and a ramet connected by a rhizome to a neighbour growing in salt-free soil in the other compartment. Ramets in the saline compartment grew more than twice as well when connected with a salt-free neighbour than when connected with a saline neighbour. The average combined performance of connected ramets growing in different environments was better than the average performance of plants growing in the two kinds of environment, connected to neighbours in a like environment. This surprising result suggests that physiological integration could cause some clonal plants to perform better in heterogeneous environments than in uniform ones.

Langer, Ryle and Jewiss (1964) studied the ramet (tiller) and genet dynamics of two grass species, timothy (*Phleum pratense*) and meadow fescue (*Festuca pratensis*). *P. pratense* is a species in which there appears to be little translocation between tillers (Table 6.1). Large concrete containers were separately sown in August with seeds of *P. pratense* to give a seedling density of approximately 4300 m^{-2} and with seeds of *F. pratensis* to give a density of about half this. The swards were cut periodically and the survival of plants and the births and deaths of tillers were followed for 3 years.

In the first 6 months of the experiment a rapid increase in tiller density and an accelerating mortality of genets was observed for both species. By 12 months the rapid changes of the establishment phase had slowed down and a regular annual pattern of tiller birth and death developed. Most of the flux in tiller numbers and most genet deaths occurred in the period of most active growth from April to July (Fig. 6.2). Tiller recruitment and death continued at a low level during the rest of the year too, but genet deaths were almost exclusively confined to this period. Apart from the flushes of tiller recruitment and death during the growing season, overall tiller densities remained approximately constant after the establishment phase because genet deaths were offset by the vegetative propagation of tillers from the survivors. As a result of annual genet mortality and tiller production of surviving genets, the average number of tillers per plant rose over the 3 years of the experiment. Langer and his associates summed up the changes they observed in these experimental grass swards: 'From a community composed of many plants (genets), each bearing very few tillers, the swards were gradually transformed into populations of relatively few, multi-tillering plants.'

The natural colonization of an old field in Illinois by *Solidago canadensis* observed by Hartnett and Bazzaz (1985a) followed a similar pattern to that described by Langer *et al.* for experimentally sown grasses. Invasion began in the third year after abandonment (1977). The density

Fig. 6.2 (a) The number of surviving genets per square metre in *Phleum pratense* (broken line) and *Festuca pratensis* (continuous line) populations planted in experiments by Langer, Ryle and Jewiss (1964). Fig. 6.2 (b) The number of tillers per square metre in the populations of *Phleum pratense* (broken line) and *Festuca pratensis* (continuous line). (Redrawn from Langer, Ryle and Jewiss 1964)

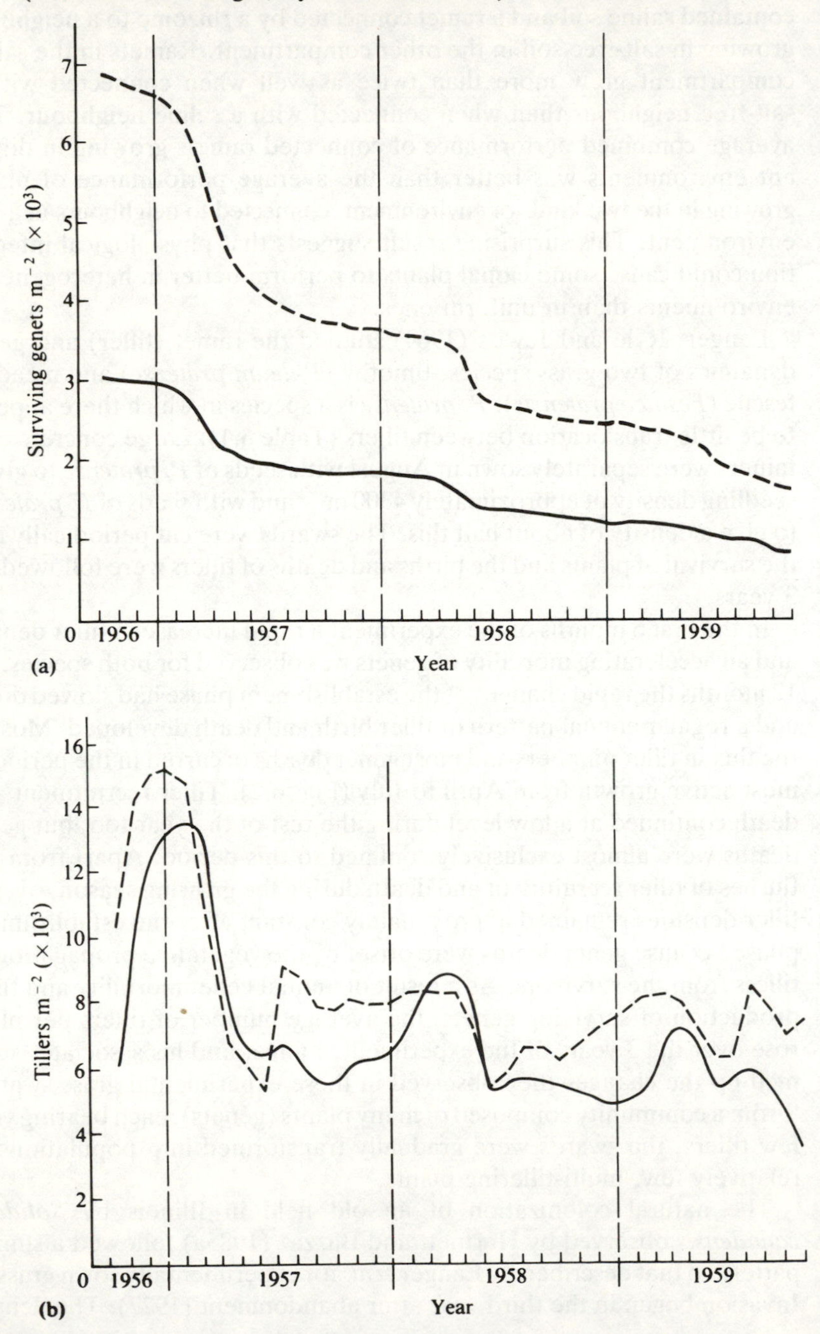

of ramets increased steadily from this time (Fig. 6.3(a)), mostly due to the spread of the first genets to colonize (Fig. 6.3(b)). Genets arriving later grew less rapidly and suffered much greater mortality.

The establishment and persistence of clonal populations

The tendency towards a reduction in the number of genets in a population, accompanied by an increase in the number of ramets per genet in the remaining plants which was observed in *Solidago canadensis*, *Phleum pratense* and *Festuca pratensis* is carried to its logical conclusion in stable habitats. Clones of sheep's fescue (*F. ovina*) up to 10 m in diameter have been found in hill pastures in Scotland (Harberd 1962). In a study of another fescue *F. rubra* which grows in the same habitat, 1481 plants were gathered within a 90 m × 90 m area and identified to individual clones on distinctive morphological characters. Most of the plants in this area belonged to one of only a few large clones. In a wider sample of *F. rubra* at this site one clone was found distributed over an area more than 200 m in diameter (Harberd 1961).

Clones of this size may be the product of centuries of vegetative propagation. Their age can be determined from rates of radial spread and the diameter of clones or, in some rare cases, from historically dated events which correlate with thé initiation of new clones. The latter method has been used to date the establishment of bracken (*Pteridium aquilinum*) clones in Finland. New clonal populations of this fern are established from gametophytes in open areas such as heathland. Although bracken spores are ubiquitous, they appear only to produce fertile gametophytes in burnt areas, possibly because burning 'sterilizes'

Fig 6.3 (a) The densities (m^{-2}) of genets (●) and ramets (○) of *Solidago canadensis* in an old field during the first 7 years after it was abandoned. (b) The mean number of ramets per genet of *Solidago canadensis* for genets establishing in 1977, 1978 and 1979. (From Hartnett and Bazzaz 1985a)

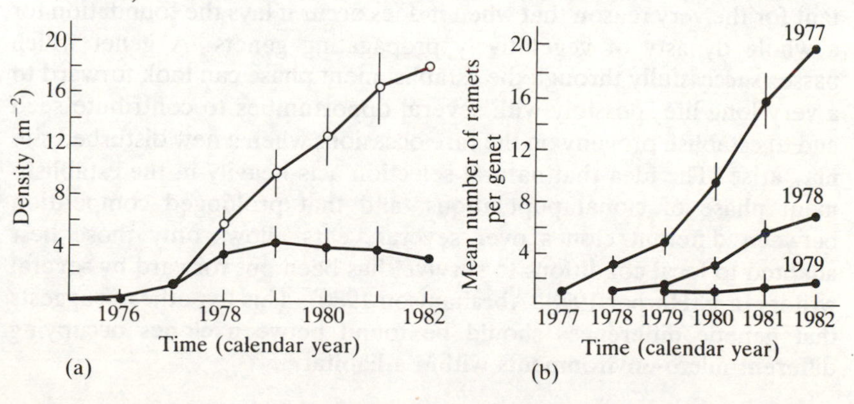

the site in some way necessary for their establishment (Oinonen 1967a). Once establishment has occurred, bracken clones are resistant to further minor fires. In an extensive survey of heathlands in Finland, Oinonen (1967a, b) identified a large number of distinct bracken clones of various sizes. In areas with small clones the date of the last fire was determined from the age of nearby trees which had also established after the fire. In areas with large clones, historical records provided dates for battles and other incendiary events.

When the date of historically or dendrochronologically dated fires is plotted against the diameter of bracken clones at the same site, a remarkably close, linear relationship is seen (Fig. 6.4). At some of the same sites where Oinonen studied bracken, he found several other clonal plants which also appear to have established after fires. The diameter of clones of the clubmoss *Lycopodium complanatum*, lily of the valley (*Convallaria majalis*) and the grass *Calamagrostis epigeios* all show a close correlation with the date of major fires (Fig. 6.4). Genets of all of these plants over 300 years old have been found.

The creosote bush (*Larrea tridentata*) is another clonal shrub, widespread and often dominant in desert areas of the Southwest USA and northern Mexico. The age structure of a population of genets expanding into a new area in southern Arizona showed a peak of 15–20-year-old shrubs with little subsequent recruitment (Chew and Chew 1965). Observations of the clonal extension of *Larrea* bushes in the Mojave desert suggest that some genets there may be thousands of years old. Central stems die and clonal extension forms a ring of shrubs (ramets) which advances radially at a rate of less than 1 mm a year. Extrapolating from this modern rate of extension, Vasek (1980) calculated that the largest clone in his study area could be 11 700 years old. An invasion of the kind observed in Arizona could therefore be a major founding event in the genetic history of a local *Larrea* population.

The spectacular ages reached by clonal plants emphasizes the rarity with which establishment from seed may occur in these species. Though extremely rare, the establishment of clonal plants from seed *is* important for the very reason that when it does occur it lays the foundation for a whole dynasty of vegetatively propagating genets. A genet which passes successfully through the establishment phase can look forward to a very long life, possibly with several opportunities to contribute seed and to establish progeny on the rare occasions when a new disturbed site may arise. The idea that natural selection acts heavily in the establishment phase of clonal populations, and that prolonged competition between different clones over several years allows only those best adapted to local conditions to survive, has been put forward by several writers (e.g. Harper 1978; Abrahamson 1980). This hypothesis suggests that genetic differences should be found between clones occupying different micro-environments within a habitat.

Fig. 6.4 The diameter of clones of *Pteridium aquilinum*, *Convallaria majalis*, *Calamagrostis epigeios* and *Lycopodium complanatum* in relation to their age, as determined from the dates of fires associated with historically recorded events. (From Oinonen 1969)

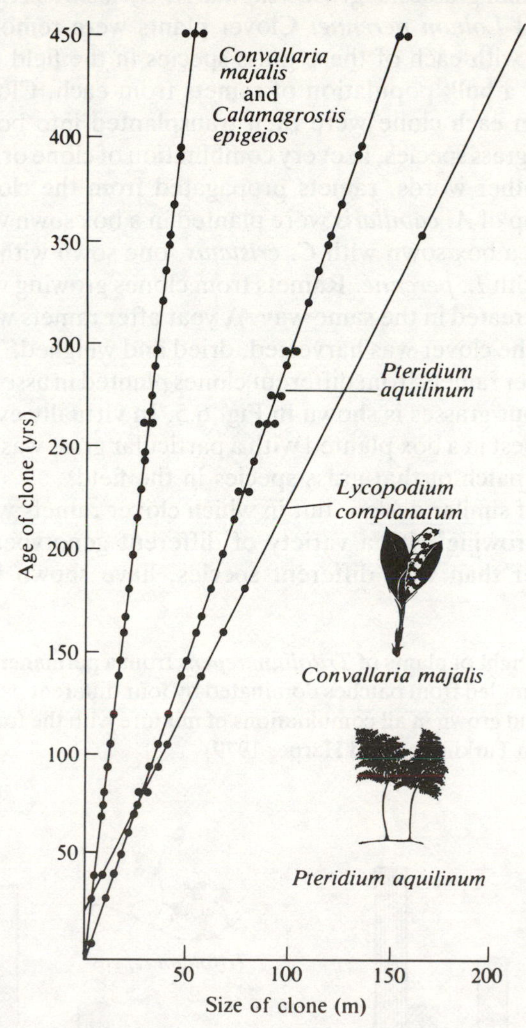

Some evidence for this was found by Harberd (1963) in a white clover population growing in sand dunes near Edinburgh. Two abundant clones, A and C, were identified and each was found to be distributed in a number of separate clumps. Clone C occurred in a number of places in a narrow zone around the rim of a dune hollow and clone A occurred at several positions within the hollow itself. In this case it appears that the extent and distribution of particular clover clones was not determined by their age but by ecological limits.

An experiment by Turkington and Harper (1979) on the same species growing in a pasture in North Wales supports this conclusion. In this particular field *Trifolium repens* is found growing in patches dominated by four different perennial grasses: *Agrostis capillaris*, *Cynosurus cristatus*, *Holcus lanatus* and *Lolium perenne*. Clover plants were removed from clones associated with each of these grass species in the field and propagated to produce a bulk population of ramets from each. Clover ramets originating from each clone were then transplanted into boxes sown with the different grass species, in every combination of clone origin and grass species. In other words, ramets propagated from the clover clone growing in a clump of *A. capillaris* were planted in a box sown with this species and also in a box sown with *C. cristatus*, one sown with *H. lanatus* and one sown with *L. perenne*. Ramets from clones growing with the other grasses were treated in the same way. A year after ramets were sown into these boxes the clover was harvested, dried and weighed. The final dry weights of clover ramets from different clones planted in association with each of the four grasses is shown in Fig. 6.5. In virtually every case the clone to grow best in a box planted with a particular grass was the clone removed from a patch of that grass species in the field.

Other experiments of similar design, but in which clover ramets were sampled from plants growing with a variety of different genotypes of *Lolium perenne*, rather than with different species, have shown that

Fig. 6.5 The dry weight of plants of *Trifolium repens* from a permanent grassland sward, sampled from patches dominated by four different perennial grasses and grown in all combinations of mixture with the four grass species. (From Turkington and Harper 1979)

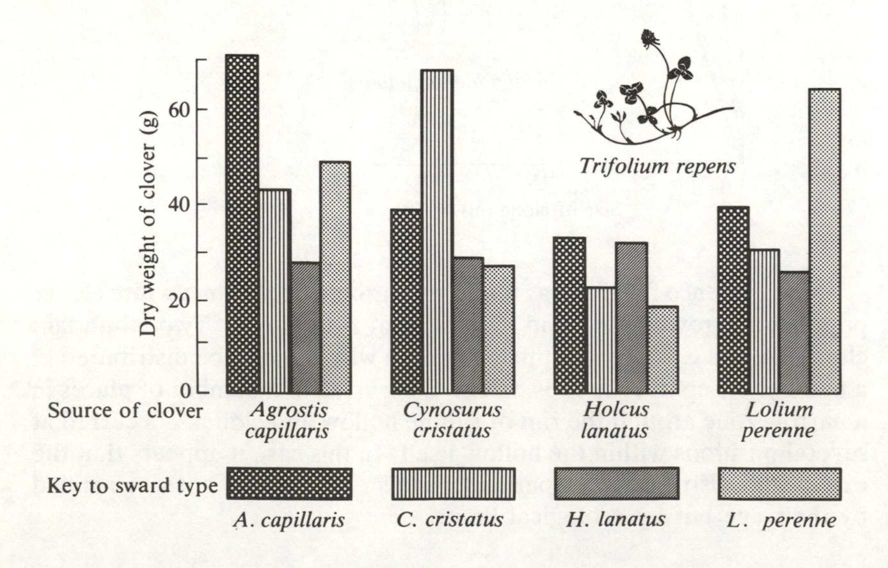

Trifolium repens is responsive to this particular grass at the sub-specific level too (Evans *et al.* 1985; Aarssen and Turkington 1985b). This strongly suggests that the identity of plant neighbours is a significant factor in determining which clover clone will successfully colonize a particular patch of ground. The stoloniferous habit confers spatial mobility and vegetative longevity on *T. repens*, both of which are probably of advantage to a clone in 'finding' (on a hit-and-miss basis) its most compatible grass neighbour and persisting in that spot once it has arrived there. This is not to be taken as a suggestion that clover plants deliberately search for a home like a bird looking for a nest site, but that ultimately competition from other clones allows each clone only to persist where it is competitively superior to the others.

Some evolutionary consequences of clonal growth

Clonal growth can lead to large, dense stands of a single genotype dominating a population. Both high density and genetic uniformity appear to increase the susceptibility of plant populations to epidemic disease and to insect attack. Pin oaks, native to North American forests but planted in monospecific rows at high density along roadsides in the suburbs of New York City, have succumbed to repeated outbreaks of a gall wasp which is killing the trees. The wasp is native to oak forests but occurs only in low numbers there (Schneider 1985). In the Netherlands 95 per cent of the elms were of one susceptible clone and some cities lost 70 per cent of their elms when elm disease appeared (Burdon and Shattock 1980).

Sexual reproduction allows organisms to produce a clutch of genetically diverse offspring. On average, these are less susceptible than the genetically uniform progeny of asexual reproduction or clonal growth because resistance to disease often has a heritable component. Thus Burdon and Marshall (1981) found, in a survey of attempts to control weeds using biological agents, that success was significantly better against asexually reproducing weeds than against sexually reproducing ones. This finding is of theoretical as well as practical interest because it supports a suggestion by Levin (1975) that sexual reproduction has an evolutionary advantage over asexual reproduction due to the protection against pathogens that genetic diversity confers.

This idea has been experimentally tested by Antonovics and Ellstrand (1984) who planted tillers of the normally outcrossing grass *Anthoxanthum odoratum* into a field in arrangements where their neighbours were clones of the same genotype or were of dissimilar genotype. They found frequency-dependent selection which favoured minority genotypes over those with clonally identical neighbours. Antonovics and Ellstrand attributed this result to the effect of pathogens, and it is interesting to note that *A. odoratum* has been shown to respond to natural selection for disease resistance in the Park Grass Experiment in England where it has evolved increased resistance to mildew attack (Davies and Snaydon 1976). There has been no systematic survey of wild clonal plants to determine whether they are more or less susceptible to disease than other plants, but studies of the clonal woodland herbs *Anemone nemorosa* (Ernst 1983), *Mercurialis perennis* (Hutchings 1983) and *Podophyllum peltatum* (Sohn and Policansky 1977) all report the occurrence of significant levels of disease in wild populations. Burdon (1980) reports that there is great variation among genotypes in disease resistance in *Trifolium repens* within the same field where Turkington and Harper (1979) studied this species.

Summary

Clonal growth is a manifestation of the modular construction of plants and is common among herbaceous perennials and shrubs. It is also found in some trees. The ramets composing clonal populations show processes of birth, death and turnover equivalent to those found among genets. However, ramets may be physiologically integrated.

In very long-lived species, the establishment of a new clone is a rare but genetically important event. Experimental observations with *Trifolium repens* suggest that the local distribution of clonal genotypes may be correlated with site characteristics, including the identity of neighbours. Other species appear to initiate new ramets in locations particularly favourable for growth.

Clones may form large stands of a single genotype. There is some evidence that high density and genetic uniformity increase susceptibility to disease but it is not known if clonal plants suffer significantly more than other species because of this.

7
Evolutionary ecology

In the foregoing chapters we have looked at the demographic characteristics of a range of plant species and have seen how variations in age-specific birth- and death-rates affect the dynamics of populations. Even though relatively few plant populations have been studied, it is clear that there are some crucial differences between species in many aspects of their life histories. Can we make sense of these differences in the reproductive and survival characteristics of species by looking at other aspects of their ecology? To what extent can the average longevity of a plant, the number of times it reproduces and the number and size of its seeds and their germination behaviour be explained as characteristics which are of some advantage in its typical habitat? In answering these questions we must look at the effects of different reproductive patterns on the fitness of plants displaying them.

Reproduction versus growth

The two fundamental components of fitness are reproduction and survival. A simple measure of these components may be obtained from the lifetime sum of age-specific fecundity multiplied by age-specific survivorship: $\Sigma l_x b_x$. The number of seeds produced by a plant, the number of seeds it fathers with the pollen it produces and the proportion of these offspring which survive to reproductive maturity are the factors which determine how many descendants are left by a genotype expressing a particular life history pattern.

It is axiomatic that natural selection favours those genotypes which leave the most descendants. Thus we may define those life history patterns which maximize the number of their surviving offspring under particular ecological conditions as *optimal* for those circumstances. Different life history patterns may be represented by life tables and fecundity schedules with different age-specific patterns of survival and fecundity. The optimal life history among a collection of possible alternatives is therefore that which produces the highest value of $\Sigma l_x b_x$. In other words this is the one with the highest fitness.

Although it might be expected that the fitness of an average individual would increase in direct proportion to the number of seeds produced, in reality reproduction incurs a 'cost' in terms of growth and survival. Plants appear to possess only limited resources which are shared between the competing demands of maintenance, growth and reproduction.

The effect of seed production on the annual growth of temperate-forest trees can be determined by comparing the size of seed crops with the width of annual rings produced in the same and following years. Resources allocated to seed production can produce a decrease in wood growth in beech (*Fagus sylvatica*) for more than 2 years succeeding a large crop (Holmsgaard 1956), though in other species such as Douglas fir (*Pseudotsuga menziesii*), grand fir (*Abies grandis*) and western white pine (*Pinus monticola*), large seed crops only reduce growth in the year of seed production itself (Fig. 7.1) (Eis, Garman and Ebel 1965). Root growth is also affected by reproduction. Even light crops of fruit have been observed to reduce root growth in apple trees.

Though we might expect growth and survival to be related, there is no direct evidence that seed production incurs a cost in fitness by increasing the probability that a tree will actually die. Such an effect has been found in some experimental populations of annual meadow grass (*Poa annua*). Despite its name, this plant frequently lives longer than 1 year and its life history varies both within and between populations. The relationship between the number of inflorescences borne by a plant at 4–5 months of age and the probability that it would survive to 18 months old was calculated for a collection of plants grown from seed under standard conditions by Law (1979). This revealed an inverse relationship between early reproduction and survival. Law also found an

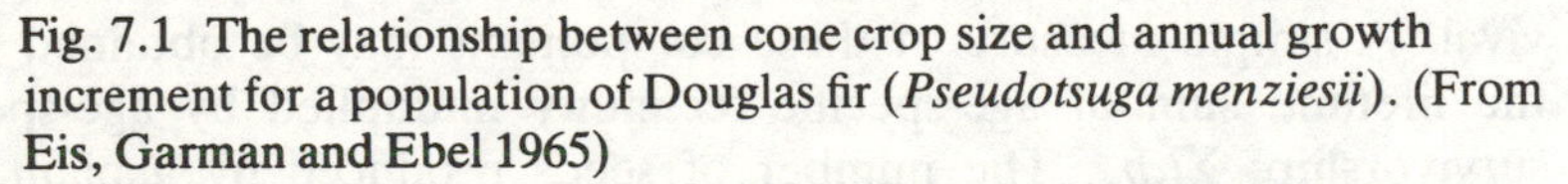

Fig. 7.1 The relationship between cone crop size and annual growth increment for a population of Douglas fir (*Pseudotsuga menziesii*). (From Eis, Garman and Ebel 1965)

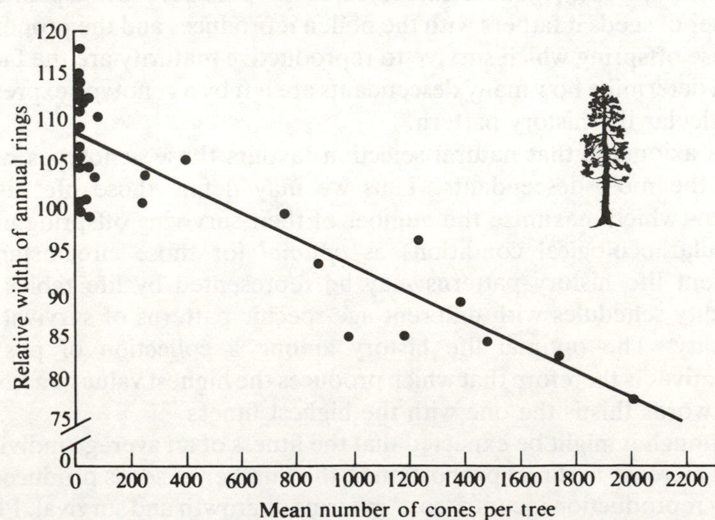

inverse relationship between the number of inflorescences produced by *P. annua* in the first season and both the size of plants and their inflorescence production in the second season.

In longer-lived herbs which spread vegetatively, seed production may exact a cost in terms of reduced clonal growth. The mayapple (*Podophyllum peltatum*) is a rhizomatous, perennial herb which occurs in deciduous woodlands of eastern North America where it was studied by Sohn and Policansky (1977). The mayapple produces both flowering shoots and new subterranean growth from its rhizome. However, sections of rhizome which produce a flowering shoot and which also bear fruit are generally incapable of producing as much vegetative spread in the following year as rhizomes which have flowers but do not produce fruit.

Mayapple clones are susceptible to various hazards, in particular to a rust disease which prevents growth altogether in the part of the rhizome attached to infected shoots. Sohn and Policansky calculated the life expectancy of model mayapple clones in which all sexual shoots bore fruit, compared with clones in which some shoots were barren. They found that the probability of eventual extinction for an entirely fertile clone was 1.0 (i.e. certain), but that the probability for a partly barren clone was 0.55. If the environmental causes of mortality did not become more severe than those measured in this study, some partially barren clones could persist indefinitely.

The dichotomous allocation of resources to reproduction or growth may not always be as clear-cut as it appears to be for *Poa* or *Podophyllum*. In many species, reproductive structures and even seeds themselves (Yakovlev and Zhukova 1980) contain chlorophyll which enables them to photosynthesize and hence to make a contribution to the energetic cost of their own production. This energetic contribution may be substantial but reproductive structures do also utilize other resources such as minerals (Bazzaz and Carlson 1979).

Evolutionary constraints

The cost of reproduction is one of the *constraints* which influences the evolution of life history characters. Structural and developmental characteristics of particular plant taxa or plant organs also constrain the direction that natural selection may cause evolution to take (Maynard Smith *et al.* 1985). For example palms have no secondary thickening in their trunks which therefore cannot support large branches. Branching in palms is rare.

Seeds from many different plant families share a similarity of basic structure: an embryo wrapped in several concentric layers of maternal tissue. This structure makes it particularly easy for an individual plant to produce seeds with different types of germination behaviour. In this case a developmental pattern opens avenues for the evolution of seed germination behaviour. Variation in the behaviour of seeds from the same plant appears to be a very widespread phenomenon (Silvertown 1984). In general, constraints of one kind or another always enter into discussions of adaptation.

The timing of reproduction and death

If reproduction and survival, which includes the chance of further reproduction, are generally alternatives to some extent, then what determines where the balance between these two options lies in a particular population? This question can be broken down into two parts. The first is: what is the optimum length of time for a plant to remain in a non-reproductive phase of growth before producing its first, and for single reproducers its last, seeds and pollen? The second question is: when does a single bout of reproduction followed by death produce a higher fitness than repeated reproduction?

Annual plants 'prove' that precocious reproduction is possible and yet many herbs delay reproductive maturity for several years. Some bamboos and some trees may even delay reproduction for several decades. In general terms, precocious reproduction should always be favoured over delayed reproduction in a growing population because of the compounding effect of reproduction. Ten offspring from an annual plant will have 10 offspring each in the following year, these will in turn have 10 offspring each and in 6 years time the original annual plant will have given rise to 10^6 descendants. A plant taking 2 years to reach maturity will have only 10^3 descendants at the end of the same period.

Several different ecological situations favour delayed reproduction, but one consideration is fundamental. Due to the vegetative costs of reproduction, a plant reproducing at small size may experience a greater risk of death. Thus $\Sigma l_x b_x$ may be maximized by delaying reproduction until the plant reaches such a size as to be able to survive, or at least to complete, its first bout of reproduction. In general, the optimum age of reproductive maturity is reached when no further increase in $\Sigma l_x b_x$ can be obtained by any further delay.

Environmental causes of mortality (not directly related to the cost of reproduction) may determine where the upper limit of this optimum age lies. In unpredictable habitats such as deserts, plants which delay reproduction experience a high risk of dying before reproductive maturity. This is a plausible explanation for the high proportion of annuals found in desert floras (Fig. 7.2).

Herbivory can also exert a selective pressure which favours early reproduction. During 6 years of observation of populations of the annual woodland herb *Impatiens pallida* in Illinois, Schemske (1978, 1984) observed that one of his populations was annually destroyed in July by a

Fig. 7.2 Percentage of the herbaceous flora accounted for by annuals plotted against coefficient of variation in total annual rainfall (CV) in five North American desert habitats. (From Schaffer and Gadgil 1975 after K. T. Harper)

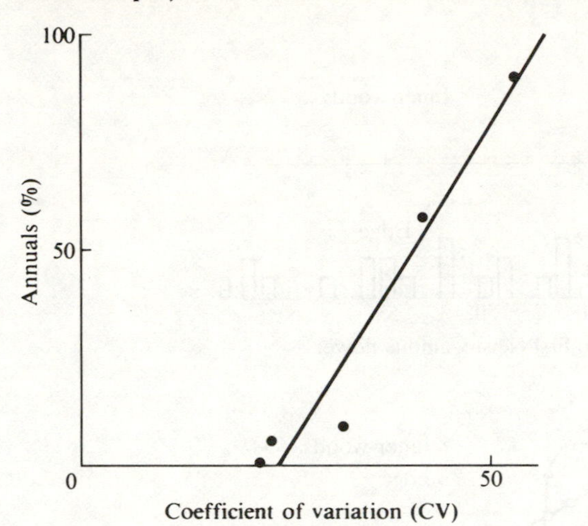

chrysomelid beetle. This population, in the interior of a wood, flowered significantly earlier than a population only 64 m away at the edge of the wood which escaped beetle attack (Fig. 7.3). In fact plants in the woodland interior which delayed flowering produced drastically fewer seeds because late-flowering plants were eaten. The flowering schedules shown in Fig. 7.3 come from greenhouse experiments with progeny derived from field populations and so these data demonstrate that differences between populations had a genetic basis (since the populations behaved differently in a common environment). Plants transplanted reciprocally between woodland interior and woodland edge had lower fitness than 'natives'. Plants from the edge did less well in the interior than natives because they were eaten before they flowered, while interior plants did less well at the edge than natives because they flowered for a shorter period of time and consequently produced fewer seeds. This study makes a convincing case that *Impatiens* was locally adapted as a result of herbivory at the woodland interior site, but in another wood where the beetle was absent, *Impatiens* showed significant genetic variation on a very local scale which was thought to be the result of genetic drift.

A high adult mortality risk favours earlier reproduction. Kadereit and Briggs (1985) found that groundsel (*Senecio vulgaris*) sampled from the flower-beds of the Cambridge Botanic Gardens developed more quickly and flowered significantly earlier than plants of the same species sampled from habitats where gardeners do not rogue out this weed.

Law, Bradshaw and Putwain (1977) compared the life histories of *Poa annua* populations collected from disturbed environments such as

Fig. 7.3 The distribution of dates of first flowering for *Impatiens pallida* collected from A, a site where it was attacked by a herbivore in July; and B, from a site where it was not attacked. C, D Survivorship of natural populations at the two sites. (Schemske 1984)

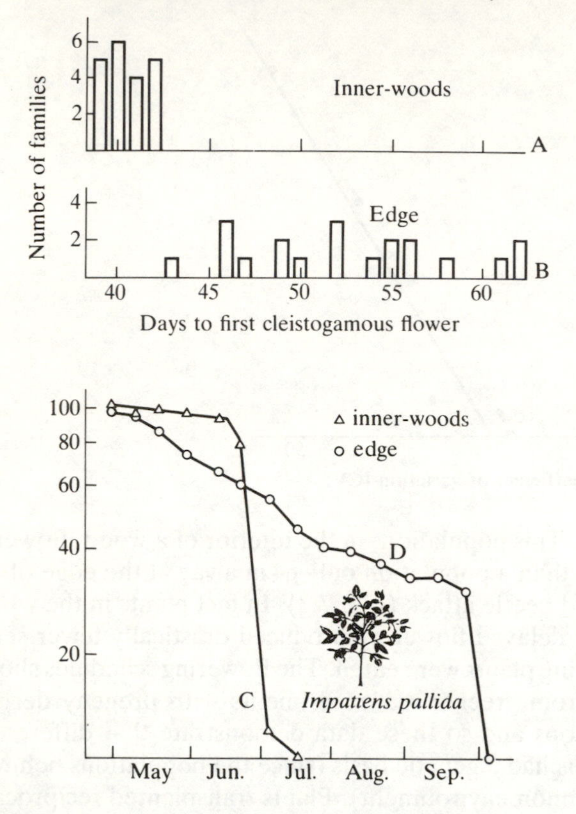

building sites where mortality risks were high, and from relatively stable pasture environments. Populations of plants grown from seeds collected in these two types of environment were raised in uniform conditions and their age-specific survival (l_x) and reproduction (b_x) were determined at monthly intervals (x) (Fig. 7.4).

The population from a disturbed environment showed early reproduction and early death, while reproduction in the pasture population was delayed so that only one major period of reproduction could be observed in the 18-month duration of the experiment. Although the short-lived population produced more inflorescences than the long-lived one in the first reproductive season (5–10 months), the situation was reversed in the second season (15–18 months). Delayed reproduction and increased survival in the pasture population allowed increased growth per plant and an increased average reproductive output.

Even if a small plant devotes all its resources to reproduction, it can

Fig. 7.4 Survivorship (dotted line) and fecundity schedules (histogram) for two populations of *Poa annua* derived from contrasting environments. (From Law, Bradshaw and Putwain 1977)

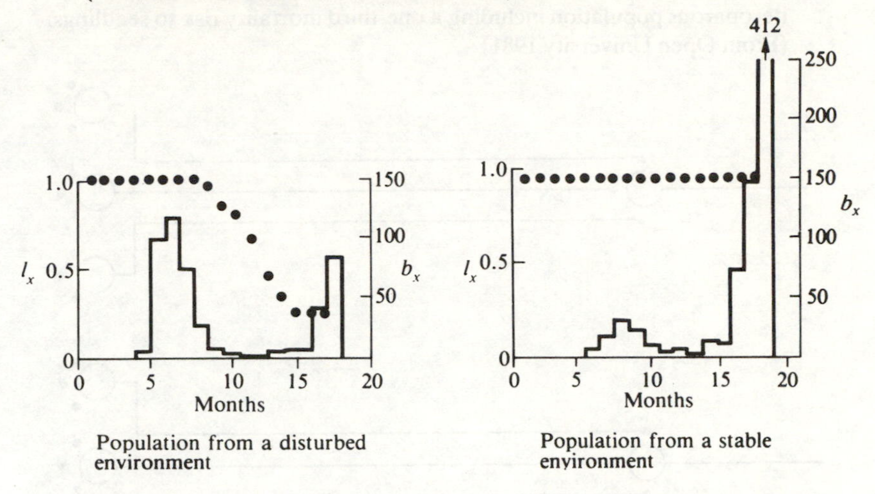

only produce a relatively small crop of seeds. Small crops generally suffer proportionately more losses to seed predators than large ones because large crops may swamp animals with more food than they can handle. This creates a situation in which a plant may gain a disproportionate release from predation by delaying reproduction until it is big enough to increase its crop size and swamp its seed predators.

This is the explanation which has been put forward for the most spectacular delays of reproductive maturity which are found in species of semelparous bamboo, among which a 20-year pre-reproductive period is common. Before they reproduce, some Indian semelparous bamboos appear to grow at an exponential rate, causing clonal genets to expand at 10 per cent a year until they flower (Gadgil and Prasad 1984). A Japanese species, *Phyllostachys bambusoides*, waits 120 years to flower and die. A few semelparous bamboos also synchronize reproduction within cohorts of the same age. This synchrony is probably important because it reduces the risk that predator populations will move from one bamboo population to another as they fruit (Janzen 1976).

Some idea of the optimal *frequency of reproduction* can also be obtained by considering the effect of fecundity and survivorship patterns on $\Sigma l_x b_x$. Consider a hypothetical population of semelparous annual plants such as the one shown in Fig. 7.5(a). Plants produce three seeds each and then die at the end of 1 year. Each seed germinates, there is no seedling mortality and nine seeds are produced in the second year for every three present in the previous one. What advantage in fitness would a mutant individual in this population gain if it became iteroparous? As Fig. 7.5(b) shows, the increased reproduction (equivalent to

Fig. 7.5 (a) A model population of three semelparous plants; (b) A model population of three iteroparous plants; (c) model semelparous population including a one-third mortality risk to seedlings; (d) model iteroparous population including a one-third mortality risk to seedlings. (From Open University 1981)

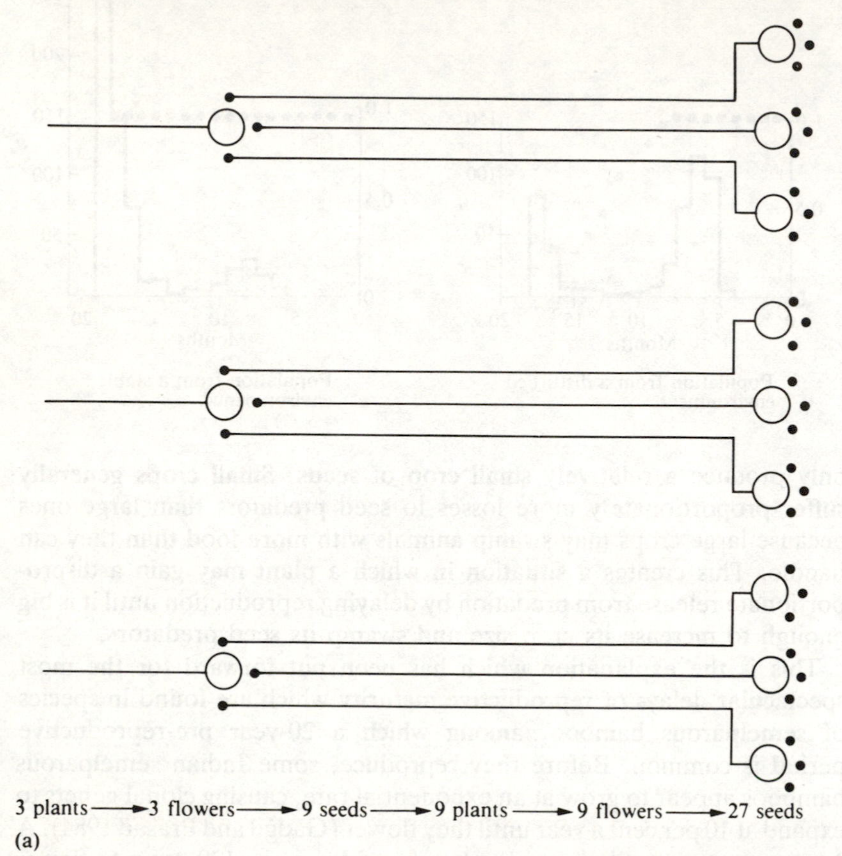

increased fitness in this case because the probability of survival = 1) of such a mutant would be equal to that of a semelparous individual which produced one extra seed before dying. It seems reasonable to suppose that the mutant would have to sacrifice more than one seed in order to switch to iteroparity in the first place so that the situation as it stands in our model appears to be heavily weighted in favour of semelparous plants. Cole (1954), who derived this result, pointed out that it paradoxically predicts that all organisms should be semelparous.

We have plainly left something important out of our model so far. In fact it is the second component of fitness – survival. Heavy seed and seedling mortality is commonplace in natural populations and values of more than 90 per cent for pre-reproductive mortality are not unusual

Fig. 7.5b, c

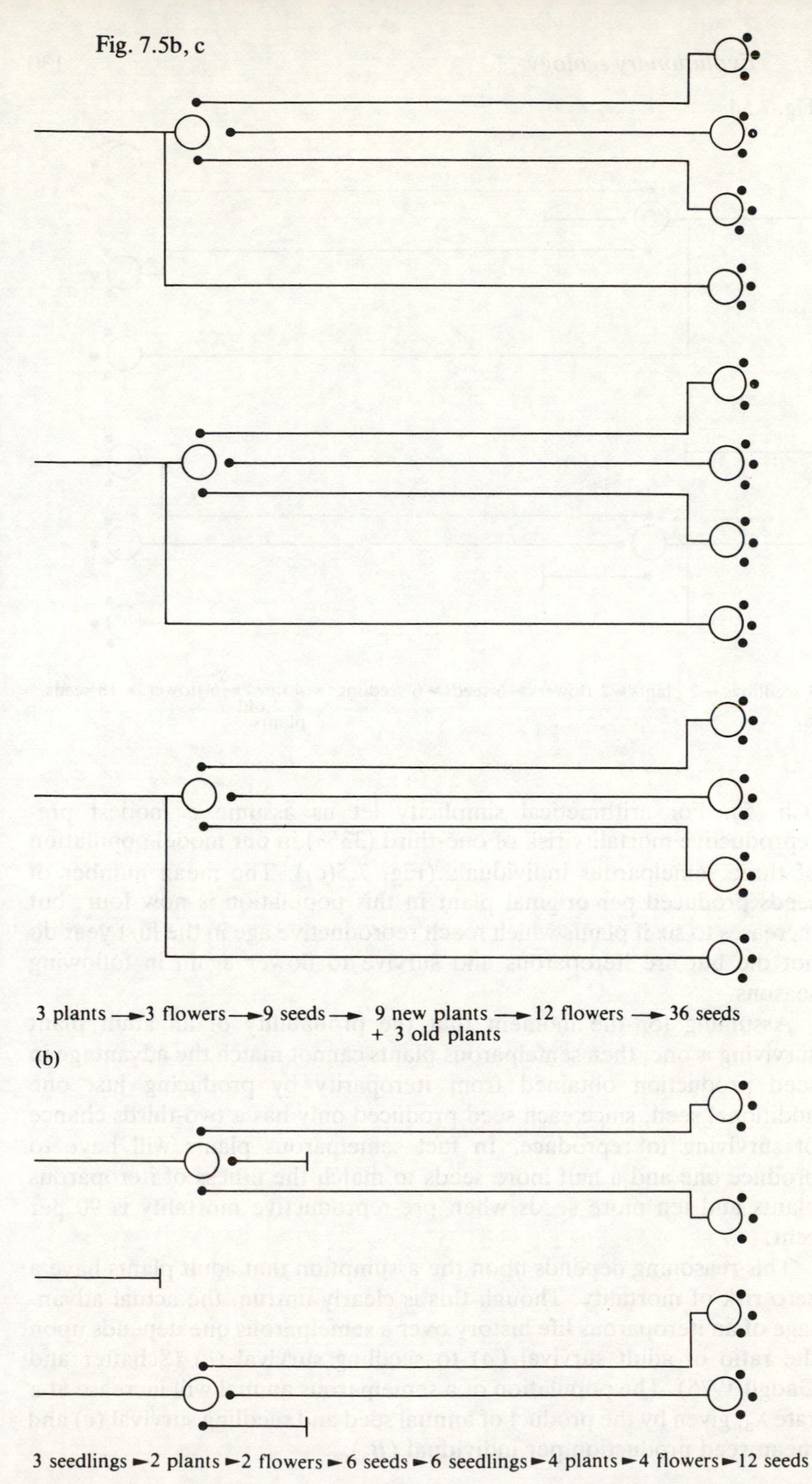
3 plants → 3 flowers → 9 seeds → 9 new plants + 3 old plants → 12 flowers → 36 seeds
(b)
3 seedlings → 2 plants → 2 flowers → 6 seeds → 6 seedlings → 4 plants → 4 flowers → 12 seeds
(c)

Fig. 7.5d

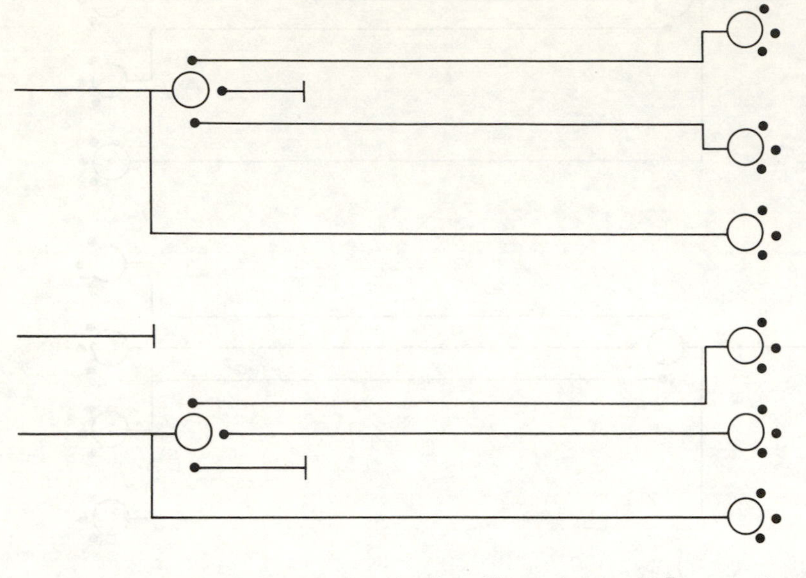

3 seedlings ► 2 plants ► 2 flowers ► 6 seeds ► 6 seedlings ► 4 new + 2 old plants ► 6 flowers ► 18 seeds

(d)

(Ch. 5). For arithmetical simplicity let us assume a modest pre-reproductive mortality risk of one-third (33%) in our model population of three semelparous individuals (Fig. 7.5(c)). The mean number of seeds produced per original plant in this population is now four, but increases to six if plants which reach reproductive age in the first year do not die but are iteroparous and survive to flower again in following seasons.

Assuming for the moment that the probability of an adult plant surviving is one, then semelparous plants cannot match the advantage in seed production obtained from iteroparity by producing just one additional seed, since each seed produced only has a two-thirds chance of surviving to reproduce. In fact semelparous plants will have to produce one and a half more seeds to match the fitness of iteroparous plants and ten more seeds when pre-reproductive mortality is 90 per cent.

This reasoning depends upon the assumption that adult plants have a zero risk of mortality. Though this is clearly untrue, the actual advantage of an iteroparous life history over a semelparous one depends upon the ratio of adult survival (p) to seedling survival (c) (Schaffer and Gadgil 1975). The population of a semelparous annual will increase at a rate λ_a, given by the product of annual seed and seedling survival (c) and mean seed production per individual (B_a):

$$\lambda_a = cB_a \quad [7.1]$$

The annual rate at which an iteroparous perennial population increases (λ_p) is given by an expression of the same form plus the mean probability of an adult surviving, p:

$$\lambda_p = cB_p + p \quad [7.2]$$

A semelparous annual will then reproduce faster than an iteroparous perennial when:

$$B_a > B_p + \frac{p}{c} \quad [7.3]$$

Sarukhán and Harper (1973) determined values of c, B_p and p for the iteroparous perennial *Ranunculus bulbosus* in the study discussed in Chapter 5 ($c = 0.05$, $B_p \simeq 30$, $p \simeq 0.8$). A semelparous mutant in this population would have to increase its seed production by 53 per cent in order to match the fitness of iteroparous plants (Schaffer and Gadgil 1975).

The survivorship curves for semelparous and iteroparous populations reviewed in Chapter 5 can now be compared with the predictions of this model. The model predicts that populations which experience relatively heavy juvenile mortality should be iteroparous, while populations with shallower curves should tend towards semelparity. The data of Figs. 5.1, 5.4 (semelparous species) and 5.6, 5.11, 5.18 (iteroparous species) are not inconsistent with this.

Using the same kind of reasoning with which we constructed [7.1] and [7.2] we can estimate the fitness of a semelparous perennial (or 'biennial') with the equation:

$$\lambda_n = (c_n B_n)^{1/n} \quad [7.4]$$

where λ_n is the yearly rate of increase of a semelparous plant living n years before flowering, c_n is the probability of survival until year n and B_n is the seed production of the plant. Notice that [7.4] is virtually the same equation as that for the annual [7.1], but raised to the power $1/n$. An annual produces cB_a offspring every year but the semelparous perennial produces its $c_n B_n$ offspring every n years. cB_a offspring produced by the annual become $(cB_a)^n$ descendants after n years (Fig. 7.5(a)), so to measure the rate of increase of the semelparous plant on the same *per year* basis as the annual we must take the nth root. Equation [7.4] is equivalent to [7.1] when $n = 1$. To compare the fitness of an annual and a semelparous perennial living two years ($n = 2$) let us conservatively assume that $cB_a = c_n B_n = 16$. The equations then show us that the relative fitness of the two life history types is 16:4 or, 1:0.25 in favour of the annual. If the semelparous perennial lives 3 years and the other variables keep the same values, relative fitnesses are 16:2 or 1:0.125.

Quite obviously delaying reproduction causes a serious decrease in fitness for a semelparous plant unless it has substantially higher seedling survival and/or much higher seed production than an annual (Hart 1977). In fact semelparous perennials do tend to have higher seed production than annuals (Salisbury 1942) but this is not the whole story. The advantage of the annual life history depends upon seeds being able to germinate every year so that its high potential geometric rate of increase (p. 124) can be realized. Conditions in many habitats make this impossible because perennial plants tend to fill in the vegetation gaps which annuals require to establish successfully (Ch. 5). An annual that has to wait for x years for a gap, with its seeds lying dormant in the soil, will have a per year rate of increase:

$$\lambda_a = (cB_a)^{1/x} \qquad [7.5]$$

Notice that the annual is now in the same position as a biennial [7.4] but with the disadvantage that it can only turn one season's assimilates into seeds. A habitat in which germination is only possible in intermittent gaps does also have a disadvantage for a semelparous perennial when compared with a completely open habitat, but the disparity in fitness between this life history and the annual one narrows dramatically as x (the time between gaps) increases (Silvertown 1983, 1986). It is therefore significant that semelparous perennials are actually most abundant in those types of habitat where gaps occur intermittently. Perennial semelparity is a relatively rare kind of life history among angiosperms as a whole but, in Europe at least, it is unusually common in families such as the Compositae and Umbelliferae which have a tap-root that forms a storage organ and an indeterminate inflorescence that can produce very large numbers of seeds from stored reserves (Silvertown 1983). These simple models of life histories indicate the kinds of ecological factors and developmental constraints which may have operated in the evolution of semelparous perennial herbs.

Looked at from a different point of view, the models also tell us something more general about the circumstances in which it is advantageous for a plant to cue its reproduction when it reaches a particular size, as opposed to when it reaches a particular age. By definition, annuals reproduce at a particular age (<12 months), irrespective of how large they are. In some situations plants like *Erophila verna* (Ch. 4) will flower when they are less than 10 mm high. We have seen that this behaviour can be advantageous in open habitats but not in others where the opportunities for seedling establishment in the year after the birth of the parent annual are low or chancy; or in other words not in sites where the local habitat *deteriorates*. In these habitats, plants which cue reproduction by size lose nothing by the delay but, on the contrary, are able to grow for several seasons and hence produce more seeds when eventually they flower (Hirose 1983; Kachi and Hirose 1985). Cueing reproduction by size, rather than by age, appears to be the rule in both semelparous and iteroparous perennials (Ch. 2) which dominate these habitats.

Reproductive allocation

As we have seen, the balance between reproduction and growth in a whole lifetime can be quantified by measuring the age of reproductive maturity, age-specific seed production and survivorship. We need to make different measurements to quantify the balance which is set between reproduction and growth within a single season. This balance is expressed as the *reproductive allocation* (RA) of a plant which is usually defined as the proportion of a plant's annual assimilated resources which is devoted to reproduction. In practice it is determined by dividing the dry weight of plants into the weight of reproductive and non-reproductive parts.

An example of how assimilated carbon is portioned between various plant organs in groundsel (*Senecio vulgaris*), an annual composite, is illustrated in Fig. 7.6. Although the total net assimilation and total seed production may be decreased drastically by stress or by interference from other plants, RA is often less severely affected. Groundsel grown in a range of pot sizes responded by a sevenfold variation in total plant weight, but RA remained at about 21 per cent in all treatments (Harper and Ogden 1970). On the other hand in experiments with another annual plant, *Chamaesyce hirta*, and with an iteroparous perennial coltsfoot (*Tussilago farfara*), RA did vary significantly with plant density (Snell and Burch 1975; Ogden 1974). The addition of nutrients to the density treatments in the experiments with *Chamaesyce* ameliorated the reduction in RA caused by crowding. In plants where pollination is not complete, RA may vary with the amount of fruit set (Colosi and Cavers 1984).

Though it is conventional to measure RA in terms of the plant biomass allocated to reproductive and to non-reproductive structures, photosynthetically fixed carbon may not be the best unit in which to assess how a plant allocates available resources. If a mineral nutrient, rather than energy, limits plant growth, it would be more meaningful to measure how that nutrient is distributed between reproductive and vegetative organs of the plant. The fact that a nutrient application to dense populations of *Chamaesyce* can partially reverse the effects of density on RA strongly suggests that energy is not the only factor limiting how many seeds these plants produce.

Biomass and phosphorus (P) allocation to different organs of a semelparous perennial ('biennial') alexanders (*Smyrnium olusatrum*) was measured in low-nutrient and in control conditions by Lovett Doust (1980) (Fig. 7.7). In the control treatment P was concentrated into reproductive organs quite disproportionately to their weight. However,

Fig. 7.6 Allocation of dry weight to the component parts of plant structure in *Senecio vulgaris* through the period from seedling to fruit production. (From Harper and Odgen 1970)

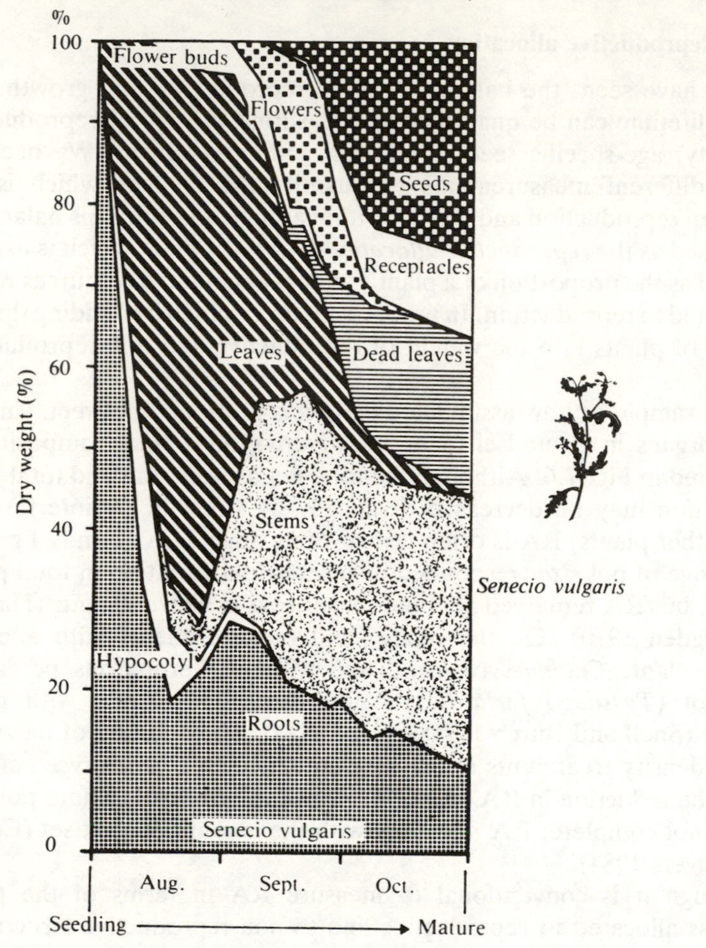

the allocations of P and biomass to reproductive structures were quite similar in the low-nutrient treatment. This suggests that P was not the limiting nutrient in impoverished soil. Selective addition of different nutrients to a low-nutrient treatment might have revealed which of the missing nutrients limited RA. Unfortunately this refinement was omitted from this experiment.

Apart from the issue of whether carbon or minerals limit RA there are also a number of other unresolved problems. Watson (1984) has suggested that in some plants, where the same meristems may give rise to either a vegetative shoot or a reproductive organ but not both, a plant's activities may be limited by the number of meristems available. For any plant of which this is true, growth and the cost of reproduction would be better measured in some morphological unit, such as meristems or nodes, than in biomass (Porter 1983a,b). Biomass is an instantaneous measure of allocation, even when plants are sampled at repeated intervals as in Fig. 7.6, and this may mislead us into thinking that a gram of reproductive tissue produced in June cost the plant the same as a gram produced in October. A fruit produced in June may actually cost more, measured in the vegetative growth that a plant sacrifices to produce it, than a fruit of the same weight produced later when the plant is larger and nearer the end of its life (Lovett Doust and Eaton 1982).

Fig. 7.7 The allocation of (a) dry matter and (b) phosphorus in control and low-nutrient treatments applied to plants of *Smyrnium olusatrum*. (From Lovett Doust 1980)

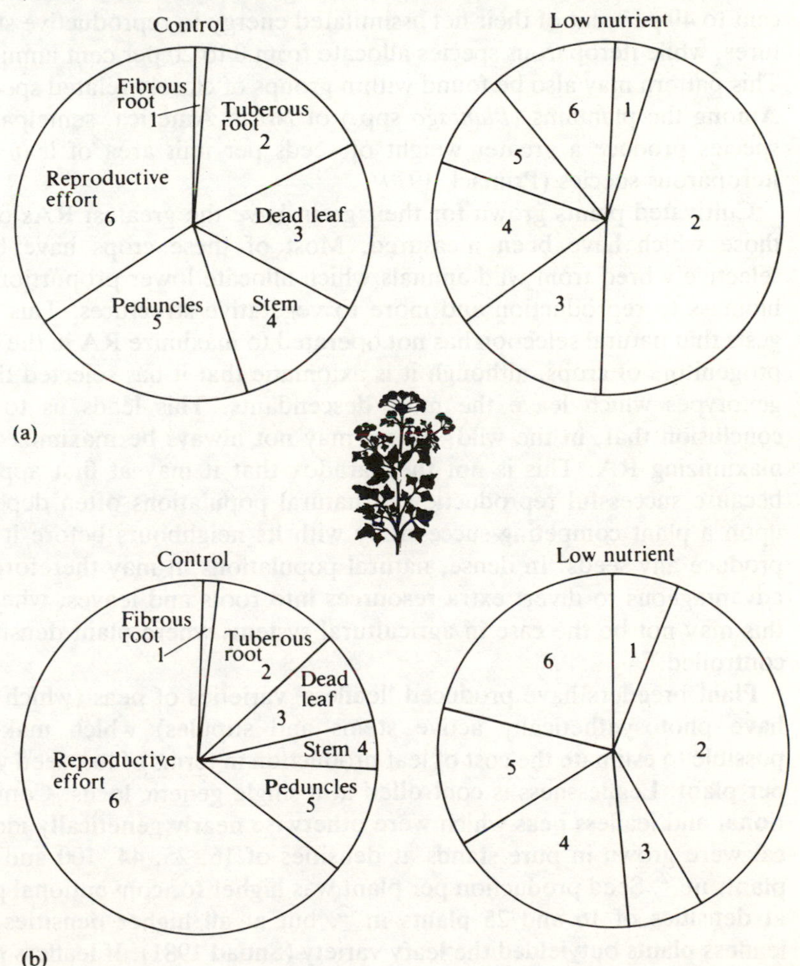

The concept of reproductive allocation is based upon the notion that reproduction and growth are competing activities in a plant. Whilst there is overwhelming evidence that this is true for most plants, there are anatomical constraints on the movement of assimilates that determine, on a local level, which reproductive organs compete with which others and which leaves supply them (Watson and Casper 1984). Vascular traces run vertically, rather than horizontally and therefore only connect leaves and flowers on the same side of a stem. Thus Alpert *et al.* (1985) found that removing the leaf subtending a flower bud of the shrub *Diplacus aurentiacus* decreased growth of the bud and increased the likelihood that it would abort. Removing the leaf opposite a flower bud had no effect upon its growth.

All these points notwithstanding, biomass is a sufficient guide to the reproductive allocation of plants for some broad generalizations to be made (see Fig. 7.8). Semelparous species generally allocate from 20 per cent to 40 per cent of their net assimilated energy to reproductive structures, while iteroparous species allocate from 0 to 20 per cent annually. This pattern may also be found within groups of closely related species. Among the plantains (*Plantago* spp.) of North America, semelparous species produce a greater weight of seeds per unit area of leaf than iteroparous species (Primack 1979).

Cultivated plants grown for their grain have the greatest RAs of all those which have been measured. Most of these crops have been selectively bred from wild annuals which allocate lower proportions of biomass to reproduction and more to vegetative structures. This suggests that natural selection has not operated to maximize RA in the wild progenitors of crops, although it is axiomatic that it has selected those genotypes which leave the most descendants. This leads us to the conclusion that, in the wild, fitness may not always be maximized by maximizing RA. This is not the paradox that it may at first appear, because successful reproduction in natural populations often depends upon a plant competing successfully with its neighbours before it can produce any seeds. In dense, natural populations, it may therefore be advantageous to divert extra resources into roots and leaves, whereas this may not be the case in agricultural systems where plant density is controlled.

Plant breeders have produced 'leafless' varieties of peas (which still have photosynthetically active stems and stipules) which make it possible to estimate the cost of leaf production in terms of lost seed yield per plant. Leaflessness is controlled at a single genetic locus. Conventional and leafless peas which were otherwise nearly genetically identical were grown in pure stands at densities of 16, 25, 44, 100 and 400 plants m^{-2}. Seed production per plant was higher for conventional peas at densities of 16 and 25 plants m^{-2}, but at all higher densities the leafless plants outyielded the leafy variety (Snoad 1981). If leafless peas

Fig. 7.8 The proportions of annual net assimilation involved in allocation to reproduction in different groups of flowering plants. (From Harper 1977, after J. Ogden)

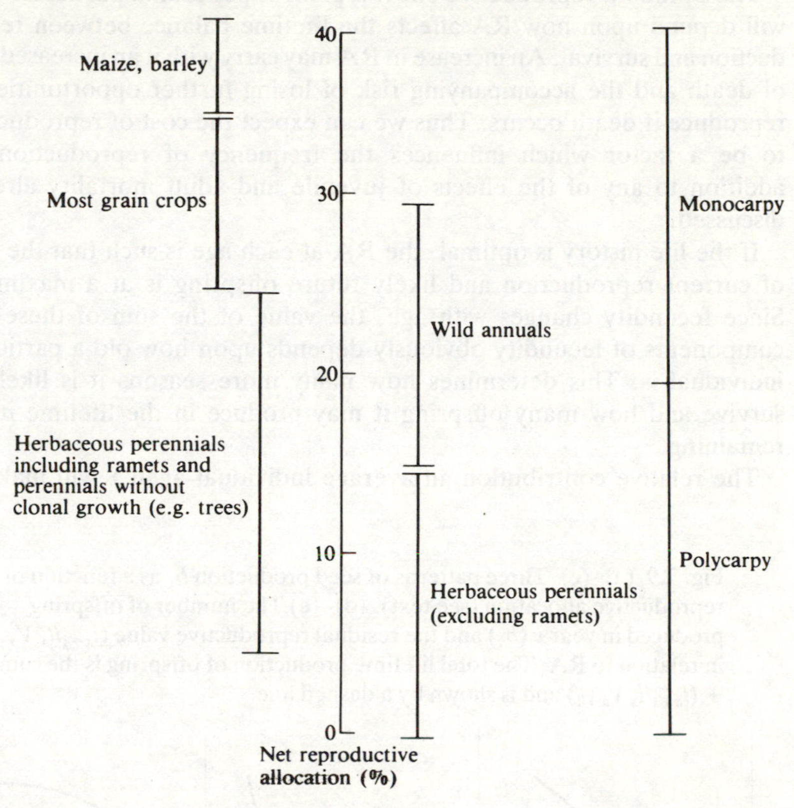

produce more seeds at high density than leafy ones, why has the leafless gene not spread through natural populations of *Pisum*? A major reason must be that such a gene would only confer increased fitness on a plant if all its neighbours were also leafless. A leafless plant in a field of leafy peas and weeds would leave few progeny!

Reproductive value

An increase in RA is not necessarily equivalent to an increase in the number of seeds produced per plant because seed production involves variable resource costs according to the size and energy content of seeds and fruits, flowers and other reproductive structures. Furthermore, density-dependent mortality among siblings may proportionally reduce the effective number of offspring produced as RA increases. The actual number of offspring produced (b_x) may vary directly with RA

(Fig. 7.9(a)), or it may increase more rapidly than RA (Fig. 7.9(b)) or less rapidly than RA (Fig. 7.9(c)).

The optimum reproductive effort a plant expends in a particular year will depend upon how RA affects the lifetime balance between reproduction and survival. An increase in RA may carry with it an increased risk of death and the accompanying risk of losing further opportunities to reproduce if death occurs. Thus we can expect the cost of reproduction to be a factor which influences the frequency of reproduction, in addition to any of the effects of juvenile and adult mortality already discussed.

If the life history is optimal, the RA at each age is such that the sum of current reproduction and likely future offspring is at a maximum. Since fecundity changes with age, the value of the sum of these two components of fecundity obviously depends upon how old a particular individual is. This determines how many more seasons it is likely to survive and how many offspring it may produce in the lifetime it has remaining.

The relative contribution an average individual aged x will make to

Fig. 7.9 (a)–(c) Three patterns of seed production b_x as a function of reproductive allocation (see text). (d)–(e) The number of offspring produced in year x (b_x) and the residual reproductive value ($l_{x+1}/l_x\ V_{x+1}$) in relation to RA. The total lifetime production of offspring is the sum b_x + ($l_{x+1}/l_x\ V_{x+1}$) and is shown by a dashed line.

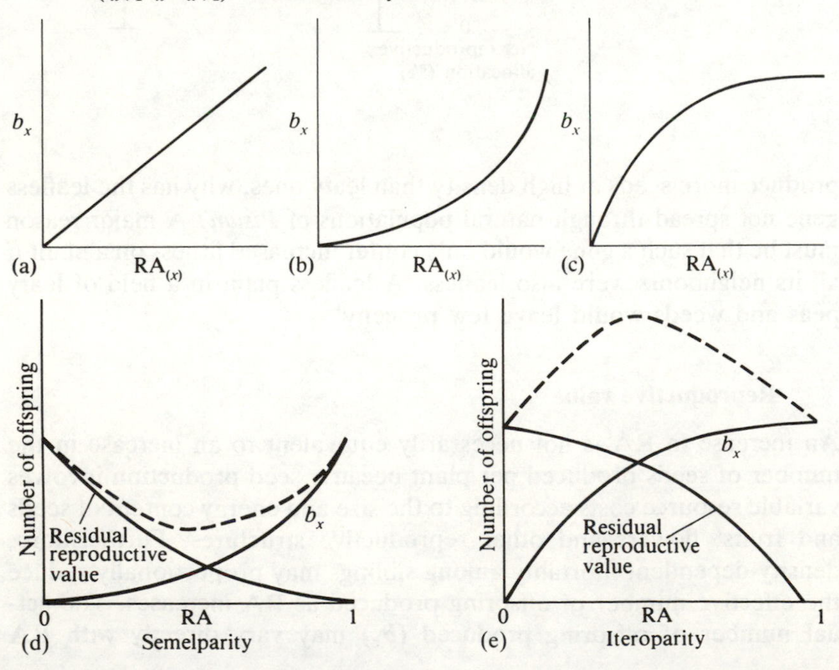

the next generation before it dies is its *reproductive value* V_x (Fisher 1930). In a stable population, V_x may be calculated as:

$$V_x = b_x + \sum_{i=1}^{i=\infty} (l_{x+i}/l_x) b_{x+i}$$

where l_x is the survivorship to age x and b_x is the average fecundity of plants aged x. These statistics are taken from the life table and fecundity schedule of a population. In words: V_x is the sum of the average number of offspring produced in the current age interval (b_x), plus the average number produced in later age intervals (b_{x+i}), allowing for the probability that an individual now of age x will survive to each of those intervals (l_{x+i}/l_x). A graph of reproductive value against age for the population of *Phlox drummondii* described in Chapter 2 (p. 13) is shown in Fig. 7.10.

The importance of the later-reproduction component depends upon the age of the plant and its *residual reproductive value*. After a plant has reproduced for the first time, its reproductive value might usually be expected to fall as it approaches the end of its lifespan. The *residual* reproductive value of a plant which reproduces in season x is equivalent to the chances which remain to it to produce further offspring in following seasons. In other words it is the probability of living one more season (l_{x+1}/l_x), times the reproductive value of a plant one season older (x + 1 seasons old) which we can denote by V_{x+1}. Therefore: residual reproductive value = $(l_{x+1}/l_x)V_{x+1}$.

While we can generally expect b_x to increase with RA, V_{x+1} will decrease because the antithesis of reproduction and growth results in a strong association between high reproductive effort and short life

Fig. 7.10 Reproductive values V_x for *Phlox drummondii*. (From Leverich and Levin 1979)

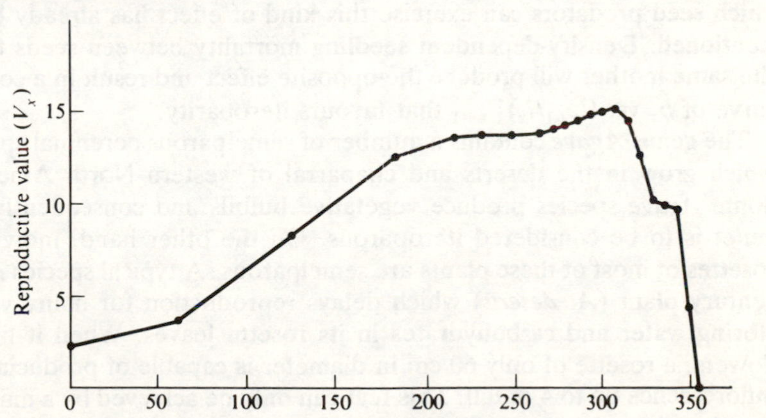

(Fig. 7.8, p. 137). The same effects of reproduction on growth and survival will also reduce the probability of a plant living an additional season as RA increases. We can calculate the total number of offspring produced for a given RA by plotting the relationship between b_x and RA and $(l_{x+1}/l_x)V_{x+1}$ and RA on the same graph (Fig. 7.9(d), (e), p. 138. The most number of offspring are produced for the value of RA which corresponds to the largest value of the sum $b_x + (l_{x+1}/l_x)V_{x+1}$.

We will assume that natural selection favours the life history which produces the largest total number of offspring in the whole lifespan of a plant. When the curves of b_x and $(l_{x+1}/l_x)V_{x+1}$ on this graph are concave (Fig. 7.9(d), p. 138), a maximum number of offspring is produced when RA = 0 or 1. In other words, if this plant is to reproduce at all (i.e. if RA is to be greater than zero), it is best for it to commit all its resources to reproduction at once instead of spreading them out. This is the semelparous life history. The iteroparous life history produces most offspring when the b_x and $(l_{x+1}/l_x)V_{x+1}$ curves are both convex (Fig. 7.9(e), p. 138).

Another way of looking at the effect of the relationship between current reproduction and residual reproductive value on the evolution of semelparity and iteroparity is to plot these two values against each other. The exact form of the inverse relationship between reproduction now and reproduction later should be different for iteroparous and semelparous populations if our reasoning so far is correct. A concave relationship is to be expected for semelparous plants (Fig. 7.11(a)) and a convex curve for iteroparous ones (Fig. 7.11(b)).

What are the ecological conditions which will result in curves of these shapes? Generally speaking, any circumstances which reduce the resource cost per seed (number of seeds per unit of RA) as the size of a seed crop increases, or which increase the probability that a seed will itself survive to reproduce as the size of the seed or seedling cohort increases, will produce a curve of the type in Fig. 7.11(a). The way in which seed predators can exercise this kind of effect has already been mentioned. Density-dependent seedling mortality between seeds from the same mother will produce the opposite effect and result in a convex curve of b_x vs. $(l_{x+1}/l_x)V_{x+1}$ that favours iteroparity.

The genus *Agave* contains a number of semelparous perennial species which grow in the deserts and chaparral of western North America. Some *Agave* species produce vegetative bulbils and consequently the genet is to be considered iteroparous. On the other hand, individual rosettes of most of these plants are semelparous. A typical species is the century plant (*A. deserti*) which delays reproduction for many years, storing water and carbohydrates in its rosette leaves. When it finally flowers, a rosette of only 60 cm in diameter is capable of producing an inflorescence up to 4 m tall. This feat can only be achieved by a massive translocation of water and assimilates from the rosette into the growing

Fig. 7.11 Two forms of the relationship between current fecundity (b_x) and residual reproductive value. In (a) the sum of the two variables is at a maximum when residual reproductive value is zero, corresponding to semelparity. In (b) the sum is at a maximum when residual reproductive value > 0, corresponding to iteroparity.

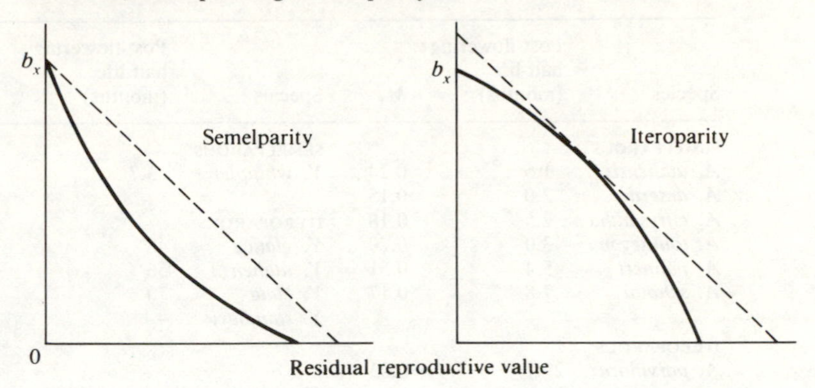

inflorescence which obtains 60 per cent of its biomass from this source.

Agaves have a similar habitat and morphology to plants in the genus *Yucca* but rosettes of these species are mostly iteroparous. Rosettes of yuccas and agaves both have stiff xeromorphic leaves and bear a central spike of insect-pollinated flowers. In an attempt to explain the difference in reproductive habit between species in these two genera, Schaffer and Schaffer (1977, 1979) compared the life history of seven *Agave* and five *Yucca* species. A sample of individual rosettes from each species was marked and the half-life of these sample populations was determined for the period after flowering. One *Yucca* (*Y. whipplei*) proved to have semelparous rosettes and one of the agaves had iteroparous ones (*A. parviflora*) (Table 4.1). This demonstrated that differences in reproductive habit between the two genera were not simply a result of different evolutionary descent, but had evolved separately in each genus.

Schaffer and Schaffer suggested that semelparity in these plants was favoured by the selective behaviour of pollinating insects which they showed visited the larger inflorescences disproportionately more often than smaller ones. They calculated the pollination advantage gained by plants with large inflorescences in different species, by plotting the number of pollinator visits observed per centimetre of flower stalk against inflorescence height. The slope (M_p) of such a graph is steep when the frequency of pollinator visits accelerates with increasing height and shallow if it does not accelerate. They then showed that the proportion of flowers which produce ripe fruit per centimetre of stalk was also related to inflorescence height and that the slope of this relationship (M_f) was correlated with M_p.

Table 7.1 Post-flowering half-life (months) and slope of the regression line of percentage of flowers developing into fruits, on inflorescence height (M_f) for twelve species of *Agave* and *Yucca*.

Agave			*Yucca*		
Species	Post-flowering half-life (months)	M_f	Species	Post-flowering half-life (months)	M_f
SEMELPAROUS			SEMELPAROUS		
A. utahensis	1.5	0.24	*Y. whipplei*	3.7	0.08
A. deserti	2.0	0.15			
A. chrysantha	2.5	0.18	ITEROPAROUS		
A. toumeyana	3.0	0.20	*Y. glauca*	48	0.00
A. palmeri	5.4	0.31	*Y. utahensis*	56	−0.02
A. schottii	7.8	0.17	*Y. elata*	70	0.01
			Y. standleyi	—	0.00
ITEROPAROUS					
A. parviflora	29.7	−0.03			

From Schaffer and Schaffer 1977

Assuming, as this evidence suggests, that pollination was the main factor determining the proportion of flowers which produced fruit, M_f was then used as a measure of pollinator preference for larger inflorescences. The relationship between M_f and post-flowering half-life for eleven species of *Yucca* and *Agave* showed that semelparous species had the highest values of M_f (Fig. 7.12). Hence these species acquired a greater benefit from increasing inflorescence height because the larger the inflorescence, the more seeds per centimetre were produced.

It may be concluded that this was probably a major evolutionary factor causing them to concentrate their reproduction in a single large inflorescence, reducing the residual reproductive value of a rosette after first flowering to zero. The iteroparous rosette species, including the iteroparous *Agave*, gained no increase in seed set per centimetre of inflorescence in larger stalks compared with smaller ones. Hence the selection pressure in favour of one large burst of reproduction was absent from these populations.

Seed size

We have so far examined, and attempted to explain in evolutionary terms, the variation to be found in the timing of reproduction and the effort expended in producing seeds. In this section we will apply the same kind of approach to variation between species in the size of seeds, and in the following two sections look at crop size and seed germination behaviour.

Fig. 7.12 Slope, M_f, of the regression line: percentage of flowers developing into fruits vs. stalk height, plotted against post-flowering half-life (PFHL) for 11 species of yucca and agave. The semelparous yucca and the iteroparous agave are marked with asterisks.

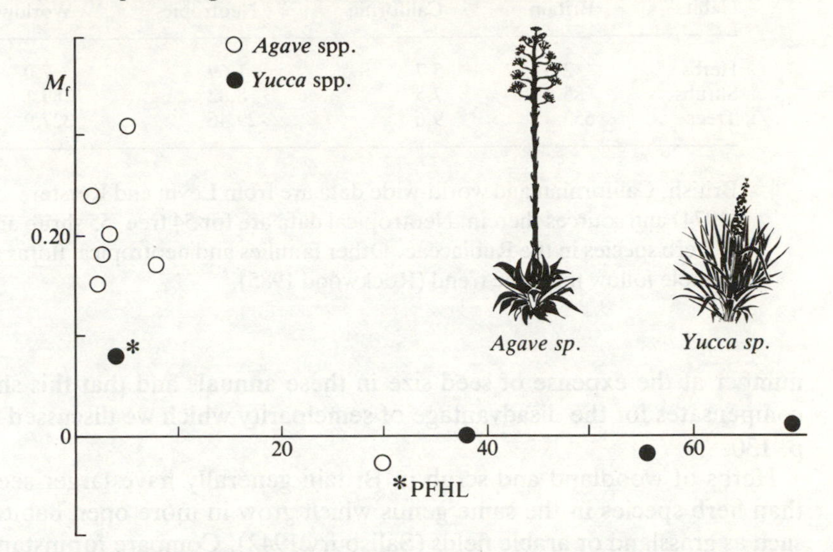

An enormous range in seed size occurs between species, from the seeds weighing 10^{-6} g produced by the orchid *Goodyera repens* to the seed of the double coconut (*Lodoicea maldivica*) which weighs in at over 10^4 g (18–27 kg) (Harper, Lovell and Moore 1970). A number of comparative studies of seed size have been made and these suggest that this character has been adjusted by natural selection in various ways, depending upon the life history and habitat of species. The mean weight of seeds increases progressively through herb, shrub and tree species in the floras of the British Isles, California and neotropical forest. It also follows the same trend on a world-wide scale (Table 7.2).

Within the herb group in California, the mean seed weight of annuals is significantly less than that of perennials (Baker 1972). This difference does not occur in Britain (Salisbury 1942; Hart 1977) when the floras of all habitats (woodland, grassland, etc.) are lumped together. Trends of seed size with variations in life history are likely to be obscured when a whole flora is compared because of the stronger association of seed size with habitat.

A comparison of seed size for annuals and perennials which occur in the same habitat enables habitat to be eliminated as a variable. When such a comparison is made for plants in the flora of British calcareous grasslands for instance, annuals are found to have significantly smaller seeds than perennial herbs (Silvertown 1981b). The annuals in this grassland sample were semelparous while most of the perennials were iteroparous. It seems likely that natural selection has increased seed

Table 7.2. Seed weight in relation to plant habit

Habit	Mean seed weight (mg) Britain	California	Neotropics	Worldwide
Herbs	2.0	5.7	1.69	7.0
Shrubs	85.4	7.5	8.82	69.1
Trees	653.4	9.6	29.46	327.9

British, Californian and world-wide data are from Levin and Kerster (1974) and sources therein. Neotropical data are for 54 tree, 55 shrub and 17 herb species in the Rubiaceae. Other families and neotropical floras as a whole follow the same trend (Rockwood 1985).

number at the expense of seed size in these annuals and that this shift compensates for the disadvantage of semelparity which we discussed on p. 130.

Herbs of woodland and scrub in Britain generally have larger seeds than herb species in the same genus which grow in more open habitats such as grassland or arable fields (Salisbury 1942). Compare for instance the seed size of *Galium anglicum*, an annual plant of open habitats in East Anglia, with that of *G. aparine* which grows in more shaded habitats and which is also annual (Fig. 7.13).

Neotropical forest trees that are able to regenerate in small gaps or in shade have significantly larger seeds than pioneer trees and those which need large light gaps (Foster and Janson 1985). This difference is presumably the result of the increased food reserves required for seedling establishment in shade, which is reflected in higher seedling mortality among species with smaller seeds in experimental shade conditions (Fig. 7.14).

In tropical forest species from Peru, Foster and Janson (1985) found that taller plants had significantly larger seeds than shorter ones of the same life form. Relationships between plant stature, shade-tolerance and seed size in modern floras have been applied to an analysis of seeds in palaeofloras by Tiffney (1984) who found that Cretaceous angiosperms had smaller seeds than those in later, Tertiary floras. On this basis, he suggests that the early flowering plants may have been mostly shrubs and small trees which occupied gaps in vegetation dominated by gymnosperms.

Variations in seed weight between species in the Californian flora appear to be more strongly related to the risk of seedling mortality due to drought than due to shade, and a positive relationship between seed weight and the dryness of the habitat occurs both among herb species within the same genus and for whole herb communities. A similar

Fig. 7.13 Fruits of two annual species of *Galium* drawn to the same scale. Above, the fruits of *G. anglicum*, a species characteristic of open situations on sandy soils, especially in East Anglia. Below a single fruit of the goosegrass (*G. aparine*), a species of woodland margins, scrub, and hedgerows. The fruits of the scrub species weigh nearly 250 times those of the open habitat species. (From Salisbury 1942)

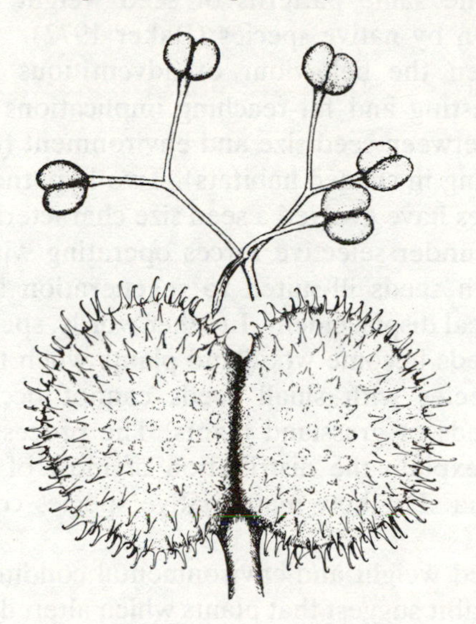

Fig. 7.14 The relation between death-rate in shade conditions and log mean seed weight in nine tree species. (From Grime and Jeffrey 1965)

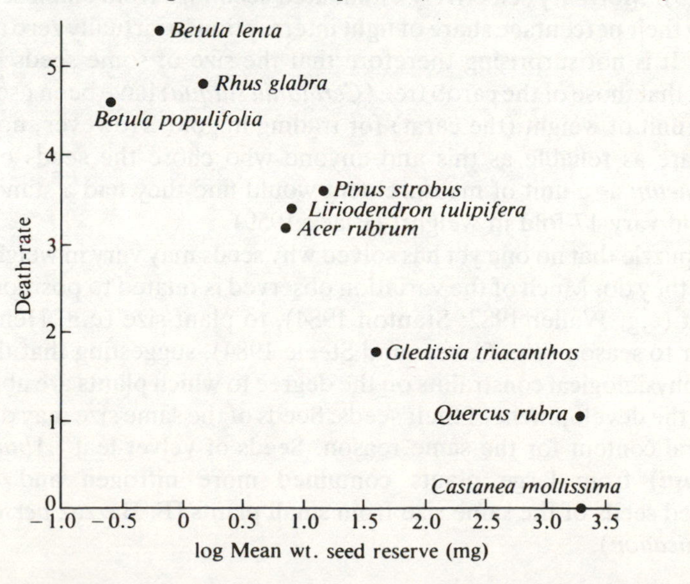

relationship between moisture availability and seed weight is also found in Californian trees. The explanation offered for these relationships is that a larger seed enables a seedling to produce a more extensive root system and thus obtain water more rapidly and efficiently than a small seed in a dry environment. California has a large adventitious flora of introduced species which fit the same patterns of seed weight and environmental conditions shown by native species (Baker 1972).

This correspondence between the behaviour of adventitious and native species has some interesting and far-reaching implications for how we view the adaptive fit between seed size and environment (e.g. species with large seeds occurring in shaded habitats). Two hypotheses for this fit are possible: 1. species have evolved a seed size characteristic of a particular habitat largely under selective forces operating within that habitat; or 2. species with seeds ill-suited to regeneration in a particular habitat suffer ecological displacement. In other words, species which happen to have large seeds become woodland plants when they invade an area. Invading species with small seeds cannot occupy woodland but can become weeds or grassland plants. The process of ecological displacement must explain the distribution of most of the adventitious plants of California. For how many native species could this also apply?

The correlations between seed weight and environmental conditions and the correlation with plant habit suggest that plants which altered the weight of their seeds beyond limits might suffer lowered fitness. The consequences of small seed size in a competitive situation are demonstrated by an experiment in which large and small seeds of subterranean clover (*Trifolium subterraneum*) were planted in a mixture (Fig. 7.15). Mortality selectively eliminated seedlings from small seeds, reducing their percentage share of light interception to virtually zero after 80 days. It is not surprising therefore that the size of some seeds is so constant that those of the carob tree (*Ceratonia siliqua*) have been used to define a unit of weight (the carat) for trading in gold. However, not all species are as reliable as this and anyone who chose the seeds of *T. subterraneum* as a unit of measurement would find they had a standard that could vary 17-fold in weight! (Black 1959).

It is a puzzle that no one yet has solved why seeds may vary in weight as much as they do. Much of the variation observed is related to position on the plant (e.g. Waller 1982; Stanton 1984), to plant size (e.g. Hendrix 1984), or to season (e.g. Cavers and Steele 1984), suggesting that there may be physiological constraints on the degree to which plants are able to canalize the development of their seeds. Seeds of the same size may differ in mineral content for the same reason. Seeds of velvet leaf (*Abutilon theophrasti*) from large plants contained more nitrogen and out-performed seeds of the same size from small plants (F. Bazzaz *personal communication*).

Fig. 7.15 The percentage of the leaf area (a) and the percentage of the light interception (b) of plants of *Trifolium subterraneum* from large and small seeds grown in a mixed sward. (From Black 1958)

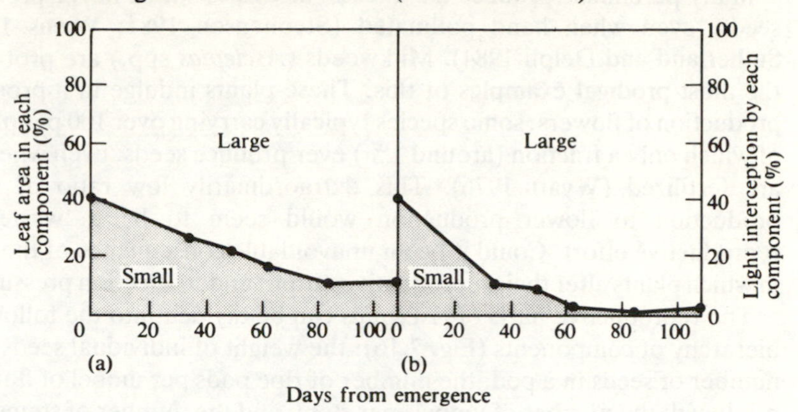

Seed crop size

The fecundity schedule of a plant describes how its lifetime seed production is apportioned and we have explored in some detail ecological situations which can favour the seemingly peculiar behaviour of semelparous plants which concentrate all their seed production into one terminal burst. But, semelparity is unusual and we also need to consider how size of seed crop is determined in other plants.

A common proximate cause of variation in seed-set between plants and of variation in the same plant between seasons is a shortage of pollinators. For example, hand-pollinated flowers of Jack-in-the-pulpit (*Arisaema triphyllum*) which is dioecious (Bierzychudek 1981), and of primrose (*Primula vulgaris*) which is self-incompatible (Piper *et al*. 1984) produced many more seeds than did naturally pollinated ones. *Primula vulgaris* has a mating system known as heterostyly, with two flower morphs that are cross-compatible but unable to fertilize their own kind. A self-compatible 'homostyle' morph occurs very locally in some populations of this species in southern England and these plants have significantly more seeds per capsule than heterostyles. Since homostyles are able to fertilize one of the heterostyle morphs as well as their own ovules, it is very surprising that homostyly has not spread in the more than 40 years that these plants have been under observation. The reason may be that there is an inverse relationship between seed size and seed number per capsule in all the primrose morphs. The two heterostyle morphs produce fewer, but significantly larger seeds than the homostyle (Boyd 1986). It is natural to think of pollinator limitation of seed-set as being a handicap, but in this case it may be that the developmental constraint which links seed size inversely with seed number turns this expectation on its head, and homostyles are unable to spread because they have smaller seeds than their pollinator-limited competitors.

Many perennials produce an 'excess' of ovules which never produce seeds, even when hand pollinated (Stephenson 1981; Wiens 1984; Sutherland and Delph 1984). Milkweeds (*Asclepias* spp.) are probably the most prodigal examples of this. These plants indulge in a prodigal production of flowers, some species typically carrying over 100 per plant, of which only a fraction (around 1%) ever produce seeds, even when all are fertilized (Wyatt 1976). This extraordinarily low ratio of seed production to flower production would seem to be a 'waste' of reproductive effort. Could it be an unavoidable consequence of the way in which plants alter their reproductive output under selection pressures?

The reproductive parts of *Asclepias* can be divided into the following hierarchy of components (Fig. 7.16): the weight of individual seeds; the number of seeds in a pod; the number of ripe pods per umbel of flowers produced; the number of umbels per stem; and the number of stems per individual plant. Wilbur (1976, 1977) compared the relative size of each of these components in seven species of milkweed which grew together in an experimental reserve in Michigan and showed that there were statistically significant differences within the genus in the way in which a given quantity of seed production was packaged. RA as defined earlier in this chapter was not measured in this study so we will use the term *reproductive output* to refer to the absolute quantity of seeds produced.

Asclepias tuberosa, *A. verticillata*, *A. purpurescens* and *A. exaltata* produced similar numbers of seeds per stem but these were packaged in large numbers in a few pods in *A. exaltata* and in smaller numbers in more pods in the other three species (Fig. 7.16). If we compare *A. exaltata* and *A. purpurescens* which produce similar numbers of seeds per plant we find that the total weight of seeds per pod is also about the same in these two species (436 and 425 mg respectively), but that this is made up of small seeds in *A. purpurescens* and larger ones in *A. exaltata*. Does this mean that seed size and seed number per pod evolved as alternative forms of reproductive packaging in *Asclepias*? A glance at the pattern of reproductive packaging in *A. verticillata* (Fig. 7.16), the species with the smallest seeds (2.14 mg) of all *and* the smallest number per pod (42), removes this idea. Indeed it is very difficult to see any clear pattern at all which could explain the differences in reproductive output and reproductive packaging in this group of species.

The plasticity of plant growth provides plants with more options and more flexibility in altering reproductive output to meet prevailing circumstances than are open to most animal species. This plasticity is sometimes itself interpreted as a response to natural selection (Bradshaw 1965). Plasticity can be expressed in different ways according to how reproductive allocation or reproductive output is adjusted in

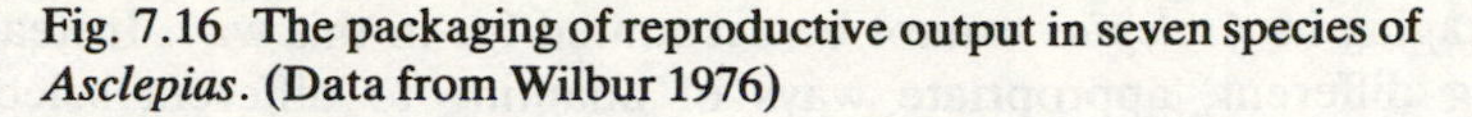

Fig. 7.16 The packaging of reproductive output in seven species of *Asclepias*. (Data from Wilbur 1976)

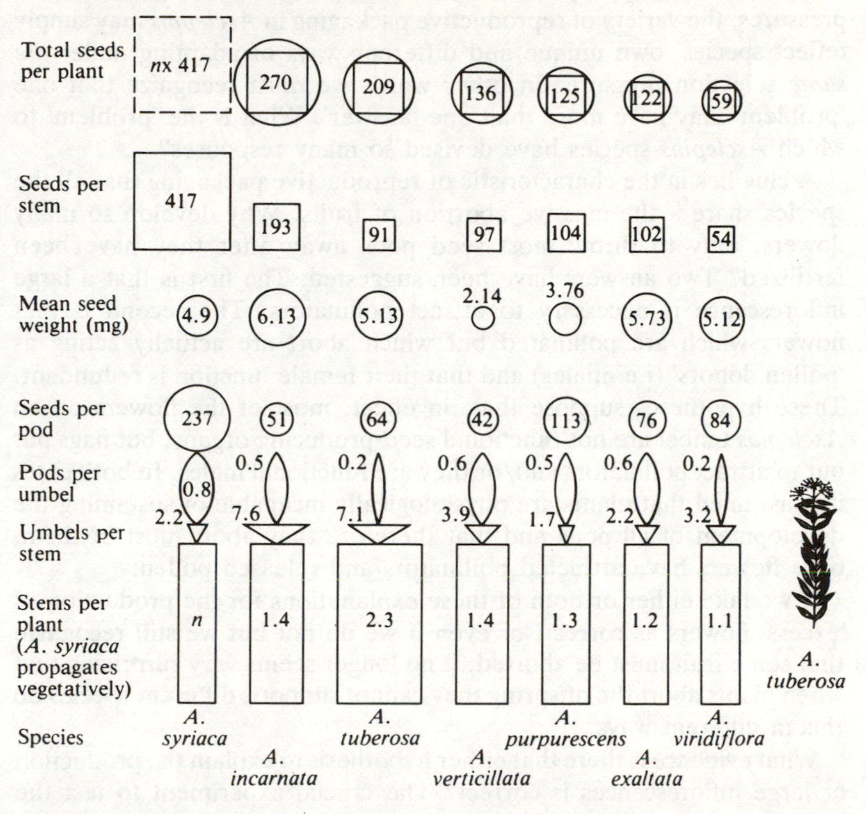

relation to energy constraints. Two different types of plastic response were shown in the asclepiads studied by Wilbur (1977). Three species, *A. incarnata*, *A. tuberosa* and *A. exaltata*, varied total reproductive output by altering the number of seeds produced through changing the number of pods aborted per umbel. However, *A. verticillata* and *A. viridiflora* altered the mean weight per seed instead of any of the other components of reproductive output, which would have produced changes in seed number. Though several different factors including predation, interference from other plants and variations in the predictability of the habitat were studied (Wilbur 1976) in an attempt to explain all these differences in *Asclepias* life history, no satisfactory 'adaptive' explanations have emerged to account for most of them.

Wilbur attempted to do what most ecologists interested in evolution try to do – to explain different patterns of life history as adaptations evolved under different selection pressures. Though this is often possible, there may be a good reason why the differences in reproductive packaging in *Asclepias* cannot be accounted for in this way. Instead of being different, appropriate ways of adapting to different selection pressures, the variety of reproductive packaging in *Asclepias* may simply reflect species' own unique and different ways of adapting under *the same* selection pressure. In other words we must recognize that one 'problem' may have more than one 'answer'. What is the 'problem' to which *Asclepias* species have devised so many responses?

A clue lies in the characteristic of reproductive packaging that all the species share – the massive abortion of fruits. Why develop so many flowers, only to throw most seed pods away after they have been fertilized? Two answers have been suggested. The first is that a large inflorescence is necessary to attract pollinators. The second is that flowers which are pollinated but which abort are actually acting as 'pollen donors' (i.e. males) and that their female function is redundant. These hypotheses suppose that, in effect, most of the flowers on an *Asclepias* umbel are not functional seed-producing organs, but flags put out to attract pollinators and/or they are functional males. In both cases it is assumed that plants are physiologically incapable of sustaining the development of all pods and that therefore they abort most of them, once flowers have attracted pollinators and released pollen.

If we take either or both of these explanations for the production of 'excess' flowers as correct, or even if we do not but we still recognize that some fruit must be aborted, it no longer seems very surprising that when plants abort the offspring they cannot support, different species do this in different ways.

What evidence is there that either hypothesis to explain the production of large inflorescences is correct? The crucial experiment to test the pollinator attraction hypothesis is to remove some flowers from umbels and then to measure pollinator visits and seed set per remaining flower. This experiment was done on *A. syriaca* by Willson and Rathcke (1974). They found that inflorescences with twenty flowers produced the greatest number of pods per flower but that smaller and larger inflorescences were less efficient producers of fertilized pods. A tenfold increase in flower number (20 to >200 flowers per inflorescence) produced only a four-fold increase in the number of pods. These results do not suggest that the commonest flower number per inflorescence found in Willson and Rathcke's population was that which maximized fecundity. Wyatt (1980) has also found an inverse relationship between fruit set per flower and the numbers of flowers in an inflorescence in *A. tuberosa*.

Asclepias flowers carry their pollen in sticky masses called pollinaria which adhere to visiting insects and are removed by them. *A. exaltata* has five pollinaria per flower and Queller (1983) used inflorescences of this species to test the 'pollen donor' hypothesis by counting the number of pollinaria removed from flowers in umbels that he had thinned to half size and in normal-sized ones. He found no difference in the number of fruits set on the two kinds of umbel, but twice as many pollinaria were removed from normal umbels as from thinned ones, demonstrating that the flowers which set no fruit do function as males.

Although Queller's experiment demonstrates that flowers which abort pods are functional males, it does not rule out the possibility that pod abortion is selective and therefore that 'excess' flowers do serve a female function by allowing plants some choice of male mate (Willson and Burley 1983). Bookman (1984) tested this hypothesis with *Asclepias speciosa* (not a species studied by Wilbur 1976 (Fig. 7.16)).

Bookman compared the seedling vigour (% germination and size) of offspring which were produced in umbels in which all flowers received pollen from the same father with the seedling vigour of offspring produced in umbels in which different flowers were pollinated with pollen from different fathers. When abortion took place, the latter situation allowed the plant to 'select' among the flowers in an umbel that had been pollinated by different fathers.

The percentage emergence of seedlings resulting from multi-father pollination was significantly greater than emergence of seedlings resulting from single-father pollinations. One-month-old seedlings from multi-paternal pollinations were significantly larger than those produced by single-father pollination. This means that abortion selects pods fertilized by fathers which produced seeds that were more viable and seedlings that had greater vigour than the average. Whether abortion is selective in other *Asclepias* species or whether *A. speciosa* is unusual is not known.

Research on *Asclepias* has revealed evolutionary explanations for several aspects of these plants' reproductive biology, but an explanation for the differences which Wilbur described between species is still elusive. The tale is a cautionary one for the ecologist who seeks an adaptive explanation for every difference between species. These may result from chance historical events or from morphological constraints and need not always be functional (Gould and Lewontin 1979; Maynard Smith *et al.* 1985).

Differences in the way in which seeds are packaged may or may not reflect selection pressures exerted at some time in the past. Total crop size is also a variable of plant fecundity which often changes from year to year within populations. The magnitude of this variation can be very different in different species. In particular, some tree species (e.g. *Fagus* spp., *Quercus* spp., *Pinus* spp.) produce vast crops of seed (mast) in some years, but few seeds in the intervening periods between mast years. Typically, seed production by different individuals in such populations is synchronized and mast years are correlated with climatic variables. In Europe, for instance, the beech (*F. sylvatica*) may mast in the year following a hot summer but rarely in the year following a cold one.

Two hypotheses have been advanced to explain masting behaviour.

The first is simply that climatic conditions suit seed production better in some years than in others and that barren periods result because trees take time to recover from the effort of reproduction (Fig. 7.1, p. 122). The second hypothesis, prompted by the observation that masting seems to waste opportunities for reproduction, is that these lost opportunities are in some way compensated because the habit increases the fitness of individual trees which vary their seed production in this way.

The argument is that seed predators consume a large proportion of small seed crops but that they cannot consume a tree's entire crop in a mast year. Hence the probability of a seed escaping predation is greatest when crops are large. However, it would be disastrous for the tree if large crops were produced regularly because predators would simply build up their numbers from one year to the next on succeeding bumper crops. This hypothesis predicts that there should be a negative relationship between the probability of a seed being eaten and the size of the current seed crop in a masting species. This prediction is confirmed by information collected by foresters on a number of species (Silvertown 1980b). The effect of two predators on seed survival in ponderosa pine is shown in Fig. 7.17. Unfortunately, however, the argument about the adaptiveness of masting does not stop here because of the correlation between seed crop size and various climatic factors.

The problem with the climatic explanation is that it does not explain why individuals in some species exhibit the masting habit more intensely than others in the same geographical region. Therefore the adaptive hypothesis might be strengthened if it can be shown that the masting habit is most pronounced in those tree populations where seed predation is strongest.

Using data on variation in the annual seed production of trees and data on seed predation in crops of different sizes (e.g. Fig. 7.17) for a range of species, it is possible to test the idea that masting species are attacked more severely than non-masting species when they produce small seed crops. Comparing seed production and seed predation data for fifteen species in this way, Silvertown (1980b) found that five of the seven most heavily preyed-upon species showed the masting habit. Among the eight species which suffered lower seed predation, only two showed very variable seed production.

Another intriguing piece of evidence which supports the idea that masting is a defensive strategy which protects trees from seed predation, comes from a comparison of the seeding behaviour of two populations of the tropical tree *Hymenaea coubaril*. This tree occurs both on the island of Puerto Rico where one of its major insect seed predators is absent, and on mainland Costa Rica where these predators are present. The mainland tree population shows the masting habit but the island one does not. Other morphological features of *H. coubaril* fruit which

Fig. 7.17 The relationship between annual cone crop size and the probability of a seed of ponderosa pine escaping predation by (a) chalcid wasps, or (b) abert squirrels. Study (a) was done in California by Fowells and Schubert (1956), study (b) in Arizona by Larson and Schubert (1970).

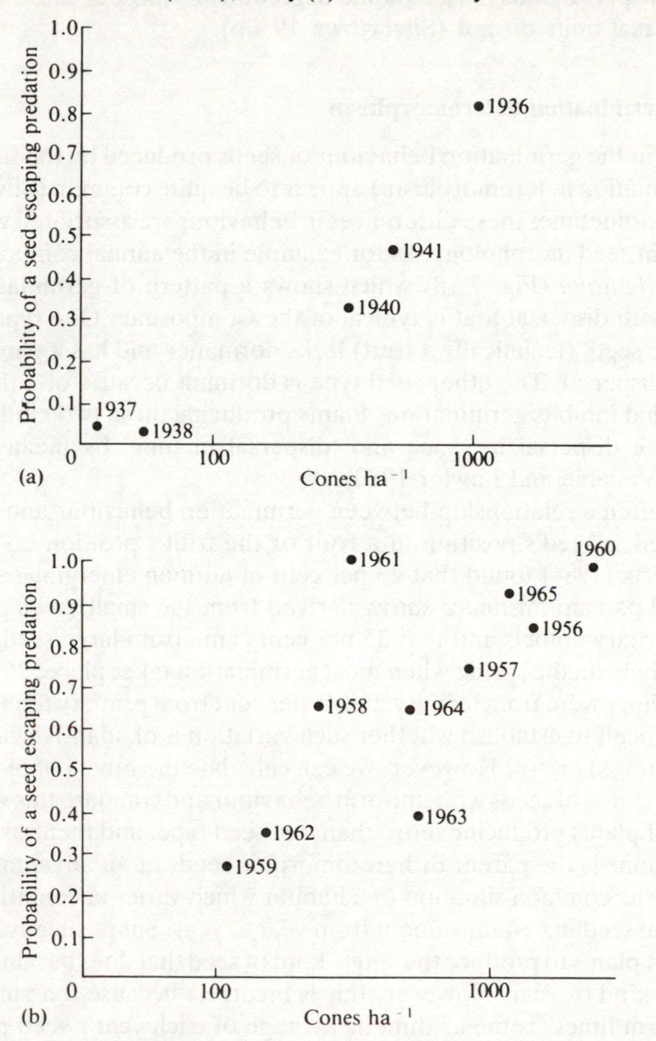

help deter predators in Costa Rica are also absent from populations in Puerto Rico (Janzen 1975b).

It follows from the argument that masting prevents animals consuming an entire tree crop that trees with fleshy fruits and animal-dispersed seeds should not mast but should produce fruit regularly. The seeds in such fruits generally pass through the gut of the dispersal agent intact, so that the only effect of masting in a fleshy-fruited species would be to prevent seed dispersal. The hypothesis that fleshy-fruited species do not mast may be tested by using a method of between-species comparison as before. A comparison of this kind, using species from the North American sylva, confirms the hypothesis and shows that most trees with non-fleshy dispersal units mast to some degree, while most of those with fleshy dispersal units do not (Silvertown 1980b).

Seed germination heteromorphism

Differences in the germination behaviour of seeds produced by the same plant (germination heteromorphism) appear to be quite common (Silvertown 1984). Sometimes these differences in behaviour are associated with differences in seed morphology, as for example in the annual composite *Heterotheca latifolia* (Fig. 7.18) which shows a pattern of germination correlated with dispersal that is typical of the Compositae. One type of *Heterotheca* 'seed' (technically a fruit) lacks dormancy and has a pappus which aids dispersal. The other seed type is dormant because of a thick fibre layer that inhibits germination. Plants producing these two kinds of seeds achieve dispersal in space and 'dispersal in time' by means of dormancy (Venable and Lawlor 1980).

There is often a relationship between germination behaviour and the size of a seed, a seed's position in a fruit or the fruit's position on the plant. Hendrix (1984) found that 45 per cent of autumn emerging seedlings of wild parsnip *Pastinaca sativa* derived from the small seeds produced on tertiary umbels and that 35 per cent came from large seeds of primary umbels. In the spring when most germination takes place, 26 per cent of seedlings were from tertiary and 45 per cent from primary umbels. It is very difficult to establish whether such variation is of adaptive value (increases fitness) or not. However, we can calculate the fitness of plants producing a clutch of seeds with uniform behaviour and compare this with the fitness of plants producing more than one seed type, and then ask: in what conditions is the parent of heteromorphic seeds at an advantage?

Consider the common situation of a habitat which varies in conditions for successful seedling establishment from year to year. Simplistically, we might expect plants to produce the single kind of seed that does best in the commonest kind of year. However, this is incorrect because it assumes that long-term fitness is the arithmetic average of each year's seed production. We have already seen that the per-year rate of increase (λ) for a population with a potential for geometric increase over several years must be obtained by taking the geometric average (p. 131) (Lacey *et al.* 1983). So here too, a geometric average is a more appropriate measure of fitness, when offspring numbers vary in time, than is an arithmetic average (Venable 1985).

Envisage two kinds of plant: a 'monomorphic producer' which produces 10 seeds of a single type (seed type I) per year and a 'dimorphic producer' which produces 5 seeds of type I and 5 of type II per year. The

Fig. 7.18 The germination of achenes derived from the disc-florets (circles) and from the ray-florets (squares) of capitula of *Heterotheca latifolia*. Open symbols are for fresh achenes, filled symbols for 2-month-old ones. The external morphology and cross-sections of the two achene types are shown inset. The pericarp of the ray achene (a) has a thick layer of fibres that inhibits germination and the disc achene (b) has a pappus which aids in dispersal. (From Venable and Levin 1985)

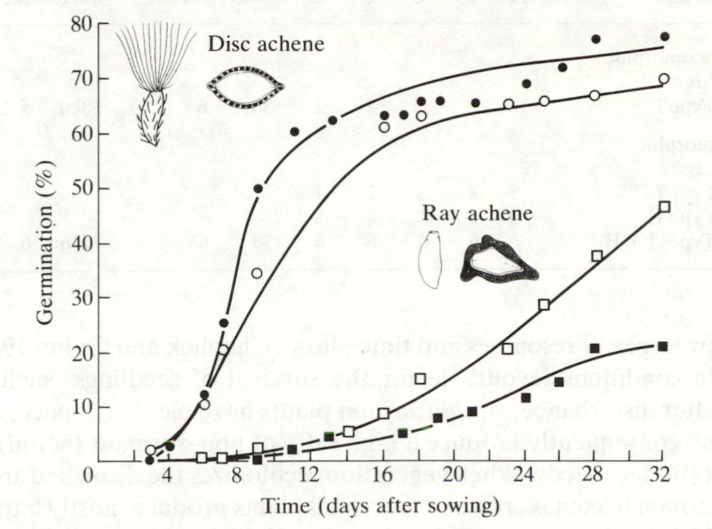

germination success of the two types of seeds is perfectly negatively correlated: i.e. when type I does well, type II does badly and vice versa. Now, compare in Table 7.3 the fitness of the two types of parent when calculated as the arithmetic mean and as the geometric mean over 5 years. Although the arithmetic calculation reveals no difference, the correct calculation using the geometric mean shows that the relative fitness of the monomorphic seed-producer is 5.2 : 6 = 0.87 : 1 and plants producing heteromorphic seeds are favoured.

In this kind of seed heteromorphism there will be a constant optimum ratio of the two seed types that maximizes the fitness of the plant, determined by the frequency of different types of year. On the other hand, when environmental conditions change in a consistent and predictable direction, the optimum ratio of seed types may change and the plant may be able to maximize its fitness by altering the ratio it produces (Venable 1985). Such changes have been observed in a number of plants which produce both aerial and subterranean seeds. This phenomenon, called amphicarpy, is rare but independently evolved in at least six different families. Peanut grass (*Amphicarpum purshii*), which occurs in sandy areas in eastern North America, is typical. New populations establish from subterranean-type seeds in disturbances. Plants produce underground seeds first and increasing numbers of aerial ones later as

Table 7.3. The number of successful offspring produced annually by plants producing the same total number of monomorphic or dimorphic seeds, and the 5-year average number of offspring of each type of parent, calculated as arithmetic and geometric means. The germination success of the two types of seed is exactly negatively correlated. (See text for further explanation).

Parental strategy	Number of successful offspring Year A	B	C	D	E	Sum	Arith. mean	Pro-duct	Geom. mean
Monomorphic seeds									
Type I	10	8	6	4	2	30	6	3840	5.2
Dimorphic seeds									
Type I	5	4	3	2	1				
Type II	1	2	3	4	5				
Types I + II	6	6	6	6	6	30	6	7776	6

they grow larger, if resources and time allow (Cheplick and Quinn 1982, 1983). In conditions favourable for the survival of seedlings, such as shortly after disturbance, *Amphicarpum* plants have plenty of space, are large, and consequently produce a high ratio of non-dormant (aerial) to dormant (buried) seeds. When vegetation recolonizes the disturbed area, *Amphicarpum* becomes crowded and small plants produce mostly buried seeds which lie dormant until the next disturbance.

To evaluate models of seed heteromorphism properly, we need to know how the success of each seed type is correlated with that of the other, and how often the conditions which favour one type or the other actually occur. This is difficult, but it can be done.

Summary

Individual plant *fitness* is determined by reproduction and survival. The lifetime sum of these two components may be expressed as $\Sigma l_x b_x$. Reproduction and growth are alternative ways in which a plant may use limited resources. The relative allocation of these resources throughout the life-cycle affects fitness.

Semelparity is favoured by a high ratio of juvenile/adult survival while *iteroparity* is favoured by a low ratio of these parameters. High mortality favours early reproduction. *Annuals* have an advantage over *semelparous perennials* ('biennials') in environments where seedling establishment is possible every year. This advantage diminishes as the interval between the appearance of vegetation gaps increases.

Reproductive allocation (RA) describes the allocation of resources to sexual reproduction within 1 year. Carbon is the resource usually used to measure this, but may be inappropriate in many situations.

Reproductive value is a measure of the number of successful offspring likely to be produced by an individual of given age before it dies.

Seed size shows correlations with plant habit (herb/shrub/tree) and with environment. Seed weight may vary within plants, perhaps because there are physiological constraints which limit the control a plant has over seed development.

Seed crop size may be limited by a shortage of pollinators or of resources. The total fecundity of a plant is determined by a hierarchy of components which may be adjusted in different ways in different species, sometimes to achieve similar results. Extreme variation in crop size, or *masting*, may increase a plant's fitness because it reduces the proportion of seeds eaten by animals.

Seed germination heteromorphism appears to be common in plants. In the Compositae many plants produce a combination of a dormant seed type lacking a pappus and a non-dormant type which has a pappus that aids dispersal. Dormancy may be regarded as 'dispersal in time'. Plants producing heteromorphic seeds may be evolutionarily favoured over those not doing so, if there is a negative correlation between the establishment success of the different morphs.

8
Interactions in mixtures of species

Despite the tendency for clonal plants to form stands of a single species, or even of a single genet, and despite the widespread practice of farmers and foresters in Europe and North America who attempt to cultivate monocultures, most vegetation in the temperate and tropical regions contains a mixture of species. No science of plant population dynamics would therefore be complete if it could not take interactions between components in these mixtures into account. We must have answers to such questions as: when can two species occupy the same habitat without one displacing the other? How does the presence of species A affect the growth and yield of species B? How does B affect A? And how do such effects change with the density and proportions of the species? Such questions on plant competition and coexistence are not easily answered by observation alone, and our most reliable answers come from greenhouse and field experiments, though the two kinds of experiment can lead to different conclusions!

Definitions of competition

Nearly every author on the subject who bothers with an explicit definition of interspecific competition has their own subtly different version, but essentially these definitions fall into two types: those which define the interaction on the basis of its *mechanism*, for example Grime (1979, p. 8): 'The tendency of neighbouring plants to utilize the same quantum of light, ion of mineral nutrient, molecule of water, or volume of space,' and those which emphasize the *outcome* of the interaction between two competing species which must, for example: '. . . cause *demonstrable* reductions in *each other's* fitness.' (Begon and Mortimer 1981, p. 52). Note that both these definitions imply that both competing species are affected (i.e. the interaction is reciprocal) though not necessarily equally (i.e. the interaction need not be symmetrical).

The second definition is the more appropriate one in the context of population ecology because we are *ultimately* interested in how *numbers* of individuals are affected by competitive interactions. Nevertheless, we should not forget two things. Firstly, the immediate effect of competition is to depress plant *growth* and *size* (performance), not plant numbers themselves. Plant numbers only change when a change in performance leads to mortality or to changes in seed or ramet production. Secondly, the outcome of competition between species is the result of a mechanism which, as Grime points out, operates between neighbours. We will see in Chapter 9 that the spatial distribution of plants can determine whether or not competition for resources, as defined by Grime, does actually result in a change in the relative abundance of species. Only such a change would be recognized as evidence of competition by a population ecologist interested in the *outcome* of interactions.

A quantitative description of competition between two species

A simple way of picturing the outcome of an interaction between two species is to plot their numbers on a graph in which the abundance of one species is measured on each axis (Fig. 8.1). This *joint-abundance diagram* can represent all possible combinations of the abundance of two species. For example the diagonal line in Fig. 8.1(a) passes through all possible ratios of *X* and *Y* which add up to a total density of 100. The vertical line in Fig. 8.1(b) passes through all possible densities of the two species in which there is a fixed number of 50 *X*. Any outcome of competition between two species can be depicted on the joint-abundance diagram by a point for the starting ratio of the species, another point for the ratio after some period of time (say, 1 year) and a line between them to represent the *trajectory* along which the composition of the mixture changes. There is an infinite number of possible mixtures of two species in the joint-abundance diagram and it is obviously impossible to draw all the trajectories that connect each starting mixture with the point representing its outcome. The trick to representing the outcome of competition

Fig. 8.1 Joint-abundance diagrams showing the densities of two species and different designs of competition experiment. In (a) the diagonal line passes through all possible proportions of two species which form a total density of 100. The three dots represent experimental treatments in a replacement series experiment. In (b) the vertical line passes through mixtures in which there is a fixed density of 50 *X* and a variable density of *Y*. The four dots represent treatments in an additive competition experiment; (c) represents an additive series experiment.

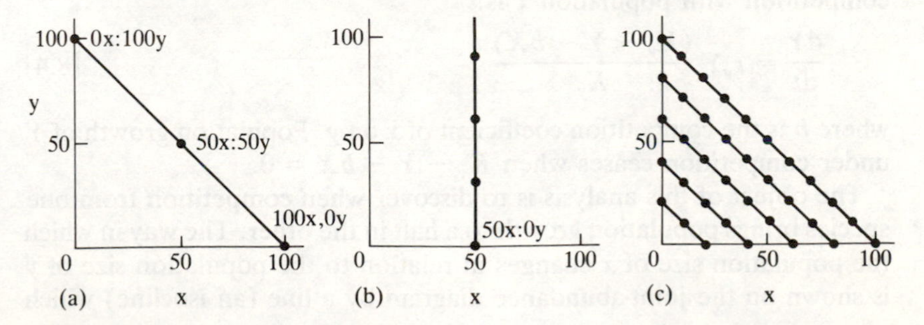

between the species, whatever the starting conditions, is to define, for each of the two species, the boundaries in the diagram which divide mixtures in which the species increases its abundance from mixtures in which it decreases. The *Lotka–Volterra equations* are a simple mathematical model of competition, based upon the logistic growth equation, which can be used to define these boundaries which are called *isoclines*. The reader should not skip over the rest of this section because it is needed to understand the design of competition experiments and will help with their interpretation.

Consider the logistic growth of a single-species population whose numbers we will measure as X. As we saw in Chapter 3 the growth of X is given by the equation:

$$\frac{dX}{dt} = r_x X \frac{(K_x - X)}{K_x} \qquad [8.1]$$

where X is the population size, r_x is the intrinsic rate of natural increase of the population and K_x is the carrying capacity of the environment for species x measured in the number of individuals of x it can support. A similar equation can be written for another species, y:

$$\frac{dY}{dt} = r_y Y \frac{(K_y - Y)}{K_y} \qquad [8.2]$$

Now imagine that species x and y are competitors and that the presence of some y individuals in the community reduces the number of x individuals that limiting resources can support, and that reduction is in proportion to the number of y individuals present, by a factor aY. The growth of population x is now given by:

$$\frac{dX}{dt} = r_x X \frac{(K_x - X - aY)}{K_x} \qquad [8.3]$$

The term aY is a measure of the interference that population y exercises on population x by reducing the carrying capacity from K_x to $K_x - aY$. The coefficient a is called the *competition coefficient*. Population growth without competition ceases when $K_x - X = 0$, and with competition when $K_x - X - aY = 0$. Since interference between populations x and y is mutual, the equivalent equation for population y when it is in competition with population x is:

$$\frac{dY}{dt} = r_y Y \frac{(K_y - Y - bX)}{K_y} \qquad [8.4]$$

where b is the competition coefficient of x on y. Population growth of Y under competition ceases when $K_y - Y - bX = 0$.

The object of this analysis is to discover when competition from one species brings population growth to a halt in the other. The way in which the population size of x changes in relation to the population size of y is shown on the joint-abundance diagram by a line (an isocline) which

Fig. 8.2 Joint abundance diagrams illustrating the outcome of interspecific competition based on the logistic model of population growth. (a) An isocline for one species *x*; (b) isoclines for two species, illustrating the competitive exclusion of *y* by *x*; (c) exclusion of *x* by *y*; (d) unstable equilibrium; (e) stable equilibrium.

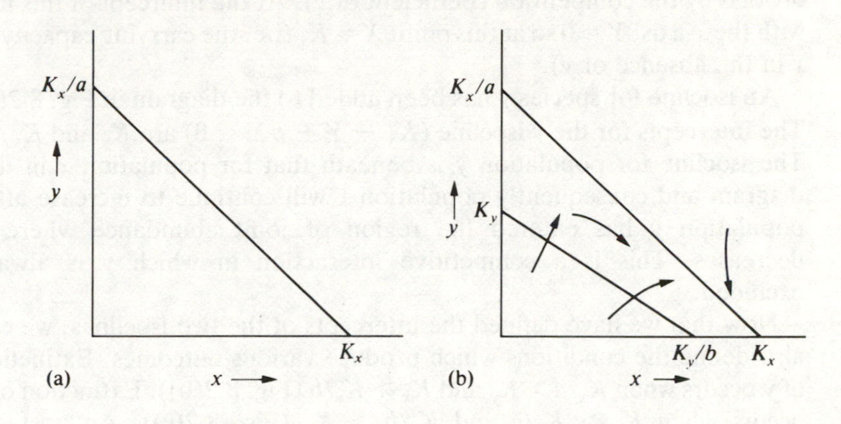

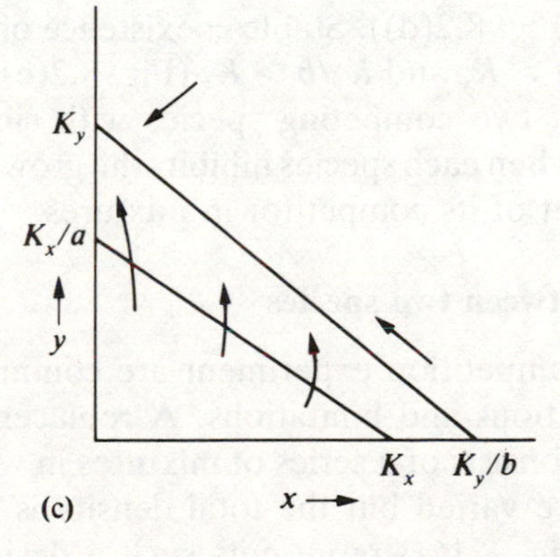

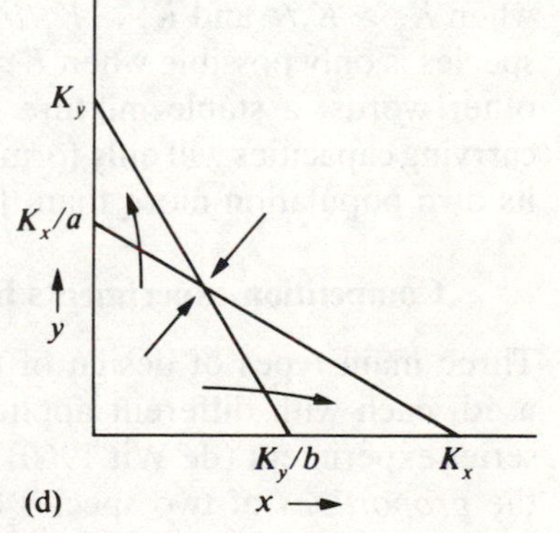

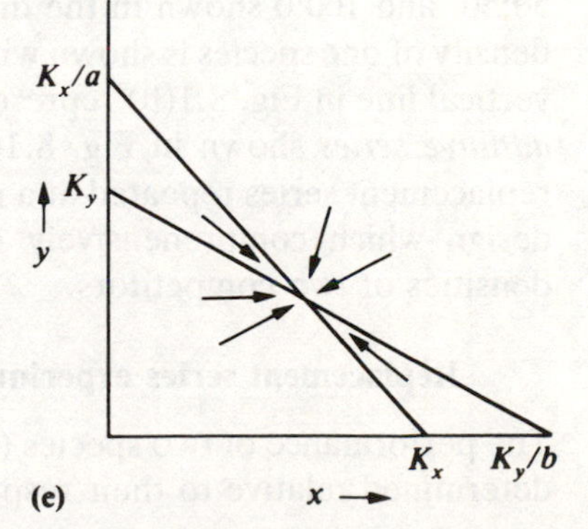

passes through all those points where the population growth of x is zero (Fig. 8.2(a)). This line corresponds to the situation in the logistic model when $K_x - X - aY = 0$. Beneath this line population X increases, above it X decreases. The intercept of the line with the y-axis is the point where $X = 0$ so the value of Y at this intercept is K_x/a (i.e. the carrying capacity of x divided by the competition coefficient of y). At the intercept of this line with the x-axis, $Y = 0$ so at this point $X = K_x$ (i.e. the carrying capacity of x in the absence of y).

An isocline for species y has been added to the diagram in Fig. 8.2(b) The intercepts for the y isocline ($K_y - Y - bX = 0$) are K_y and K_y/b. The isocline for population y is beneath that for population x in this diagram and consequently population x will continue to increase after population y has entered the region of joint abundance where it decreases. This is a competitive interaction in which y is always excluded.

Now that we have defined the intercepts of the two isoclines, we can also define the conditions which produce various outcomes. Extinction of y occurs when $K_x/a > K_y$ and $K_x > K_y/b$ (Fig. 8.2(b)). Extinction of x occurs when $K_y > K_x/a$ and $K_y/b > K_x$ (Fig. 8.2(c)). An unstable equilibrium eventually resulting in the extinction of either x or y occurs when $K_y > K_x/a$ and $K_x > K_y/b$ (Fig. 8.2(d)). Stable coexistence of two species is only possible when $K_x/a > K_y$ and $K_y/b > K_x$ (Fig. 8.2(e)). In other words, a stable mixture of two competing species with similar carrying capacities will only form when each species inhibits the growth of its own population more than that of its competitor in mixtures.

Competition experiments between two species

Three main types of design of competition experiment are commonly used, each with different applications and limitations. A replacement series experiment (de Wit 1960) consists of a series of mixtures in which the *proportions* of two species are varied but the total density is held constant. The diagonal line in Fig. 8.1(a) represents such a design – *replacement series experiments* often involve only the combinations 0:100, 50:50, and 100:0 shown in the diagram. In *additive experiments* a fixed density of one species is shown with a variety of densities of another. The vertical line in Fig. 8.1(b) represents this design. The third design is the *additive series* shown in Fig. 8.1(c). This is essentially equivalent to a replacement series repeated at a range of densities. It is the only kind of design which comprehensively explores a range of proportions and densities of two competitors.

Replacement series experiments

The performance of two species (x and y) in a replacement series can be determined relative to their respective monocultures by the formulae:

$$\text{Relative yield of } x = \frac{\text{Yield per unit area of } x \text{ in the mixture}}{\text{Yield per unit area of } x \text{ in monoculture}} \quad [8.5]$$

Similarly for y:

$$\text{Relative yield of } y = \frac{\text{Yield per unit area of } y \text{ in the mixture}}{\text{Yield per unit area of } y \text{ in monoculture}} \quad [8.6]$$

The sum of the relative yields gives the Relative Yield Total (RYT) which, under restrictive conditions and *for that particular total density only*, indicates whether the species are performing better in mixture than in monocultures. *On condition* that each component of the mixture has a constant yield per plant in pure stand across the range of single-species densities used in the experiment, a RYT > 1 indicates that a yield advantage is obtained in mixture, RYT = 1 indicates no advantage and RYT < 1 indicates a disadvantage. RYTs cannot be interpreted in this way if the condition of constant yields is not met because, without it, a mixture sown with 50 individuals of each species can produce a mixture that deviates from 50:50 proportions measured in biomass. If this is so, RYT will reflect the individual size/density responses of the species whose effect on RYT will be confounded with the effects of competition.

Because replacement series experiments are carried out at a fixed density, they cannot be used to determine how the composition or yield will behave in mixtures in which density is not held constant (Inouye and Schaffer, 1981). This is the situation in most natural vegetation and crop/weed mixtures. Replacement series experiments have been very widely used but, contrary to what many authors (including Silvertown 1982) have written, the results cannot be interpreted easily, either because the conditions mentioned above have not been satisfied or because they are so restrictive that no valid generalizations can be made which apply beyond the particular circumstance of the experiment (Jolliffe *et al*. 1984; Connolly 1986). The central problem is that the performance of plants in a replacement series mixture is determined relative to monocultures (i.e. the divisors in [8.5] and [8.6]) at the arbitrarily chosen fixed density. A choice of different monoculture densities as reference points can produce a different value of RYT and hence a different interpretation (Willey 1979a; Connolly 1986). The conclusions of most studies using this design require re-examination.

Crop mixtures and the Land Equivalent Ratio

An alternative to the Relative Yield Total which can be used to measure the performance of mixtures in relation to pure stands of crops in the *Land Equivalent Ratio* (LER) (Mead and Willey 1980). Unlike RYT, this index does not assume that crop densities will be held constant. It is used to compare the land area needed to grow any mixture of interest, planted at any total density, with the area that would be required to obtain the same yield from a series of separate monocultures of the mixtures' component species. For two species *x* and *y*, the LER is calculated as:

$$\text{LER} = \frac{\begin{array}{c}\text{Yield per unit area}\\ \text{of } x \text{ in mixture}\end{array}}{\begin{array}{c}\text{Yield per unit area}\\ \text{of } x \text{ in monoculture}\end{array}} + \frac{\begin{array}{c}\text{Yield per unit area}\\ \text{of } y \text{ in mixture}\end{array}}{\begin{array}{c}\text{Yield per unit area}\\ \text{of } y \text{ in monoculture}\end{array}}$$

where the monoculture densities are usually the ones which produce the highest yields, not necessarily those which correspond to the total densities of plants in the mixtures. An LER value > 1 indicates a yield advantage in the crop mixture. Like RYT, LER has the disadvantage that its value will change, depending upon which particular monocultures are used as standards of comparison. A method for comparing LERs which have been calculated with different standards is given by Riley (1984).

Yield advantages from intercropping can occur for a great variety of reasons including lower levels of pathogens, insect pests and weeds in intercrops and a greater use of available resources by plants with different mineral requirements (e.g. legumes and grasses), different leaf canopy structure (Fig. 8.3) or different development time to maturity. Spatial and temporal complementarity between crops often occur together, as for example in a common Mexican intercrop where early-maturing climbing beans are grown supported on the stems of later-maturing maize. Temporal complementarity appears to be very important in many intercrops with annual plants. Calculated on the basis of LERs, Willey (1979a) reports an 80 per cent advantage for an intercrop of pearl millet maturing in 85 days growing with groundnut maturing in 150 days, 30–40 per cent advantages with 90-day maize and 160-day rice, and a 38 per cent advantage with 85-day beans and 120-day maize.

Mixed cropping is the rule not the exception in peasant agriculture in the tropics. Ramakrishnan and co-workers have studied a system of slash and burn agriculture, locally known as *jhum*, in the northeastern hill region of India. Thirty or more crop species may be sown in a single plot of between 1–2.5 ha (Table 8.1). Plots which have been cropped less than 5 years before are sown with tuber crops and perennial species which are able to yield well in relatively nutrient-poor soil. Plots that have lain fallow for longer, up to 30 years, have higher soil nutrient status and are sown with cereal crops. This system of farming has many advantages over monocultures, perhaps the most important one being that it prevents soil erosion because vegetation cover is present for most of the time (Ramakrishnan 1984). This is not a factor that a one-season experimental assessment of the LERs of mixtures can take into account because the deterioration which occurs in plots sown with monocultures progresses

Fig. 8.3 Profile of a forest garden in Java showing the use of different levels in the canopy by different crops. In the ground layer (< 1.5 m) are vegetables such as spinach, beans, cucumber, tomato and medicinal plants. Between 1.5 m and 5 m are plants such as taro, cassava, banana, papaya and ornamental shrubs. In the upper levels at various heights are bamboo, coffee and cacao, fruit trees such as mango and rambuttan and sugar palm. The tallest trees are durian and coconut. (From Michon 1983)

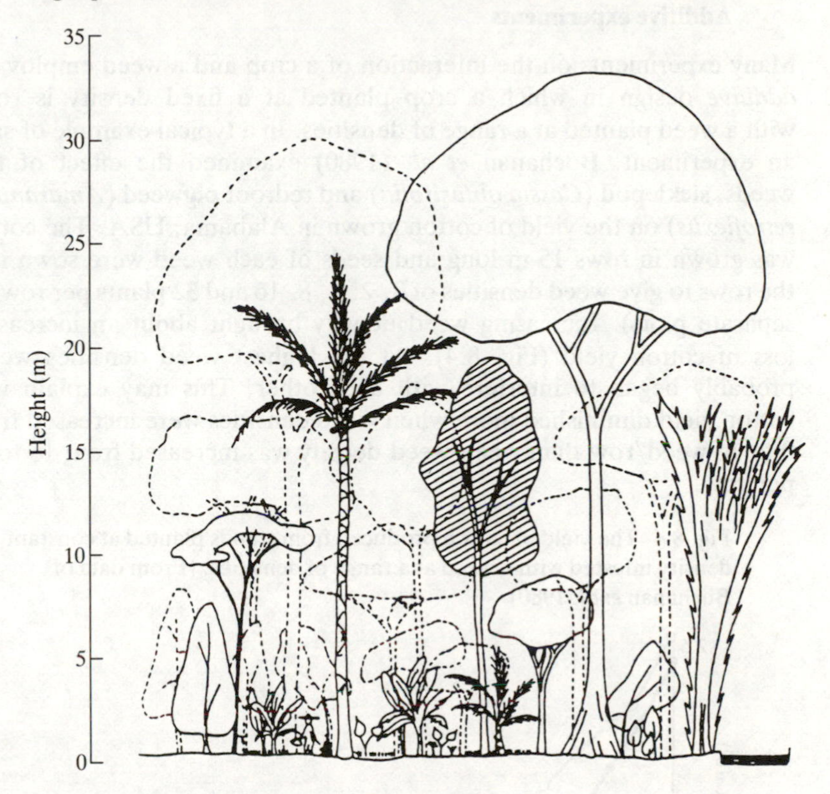

Table 8.1 Some of the crops grown in slash and burn plots at lower elevations in northeastern India (Toky and Ramakrishnan 1981)

Grain and seed		Leaf and fruit vegetables	
Oryza sativa	rice	*Hibiscus sabdariffa*	
Sesamum indicum	sesame	*H. esculentus*	okra
Zea mays	maize	*Capsicum frutescens*	chilli
Setaria italica	millet	*Lagenaria lencantha*	gourd
Phaseolus mungo	mung bean	*Cucurbita maxima*	squash
Ricinus communis	castor bean	*Cucumis sativus*	cucumber
		Momordica charantia	bitter gourd
Tubers and rhizomes		*Musa sapientum*	banana
Manihot esculenta	cassava	*Morus alba*	mulberry
Colocacia antiquorum	taro	(leaves for silkworms rearing)	
Zingiber officinale	ginger		

over several years. The complexity of this kind of multiple-crop agriculture should not be underestimated, and we shouldn't imagine that its efficiency or (more important) its ability to meet the needs of peasant farmers can be calculated solely on the basis of yield experiments with mixtures.

Additive experiments

Many experiments on the interaction of a crop and a weed employ the *additive* design in which a crop planted at a fixed density is sown with a weed planted at a range of densities. In a typical example of such an experiment, Buchanan *et al*. (1980) examined the effect of two weeds, sicklepod (*Cassia obtusifolia*) and redroot pigweed (*Amaranthus retroflexus*) on the yield of cotton grown in Alabama, USA. The cotton was grown in rows 15 m long and seeds of each weed were sown into the rows to give weed densities of 0, 2, 4, 8, 16 and 32 plants per row (in separate plots). Increasing weed density brought about an increasing loss of cotton yield (Fig. 8.4). At the highest weed densities weeds probably began to interfere with each other. This may explain why cotton yield diminished more when weed densities were increased from 0 to 16 weed/row than when weed density was increased from 16 to 32 per row.

Fig. 8.4 The yield of cotton produced from stands planted at constant density infested with a weed at a range of densities. (From data of Buchanan *et al*. 1980)

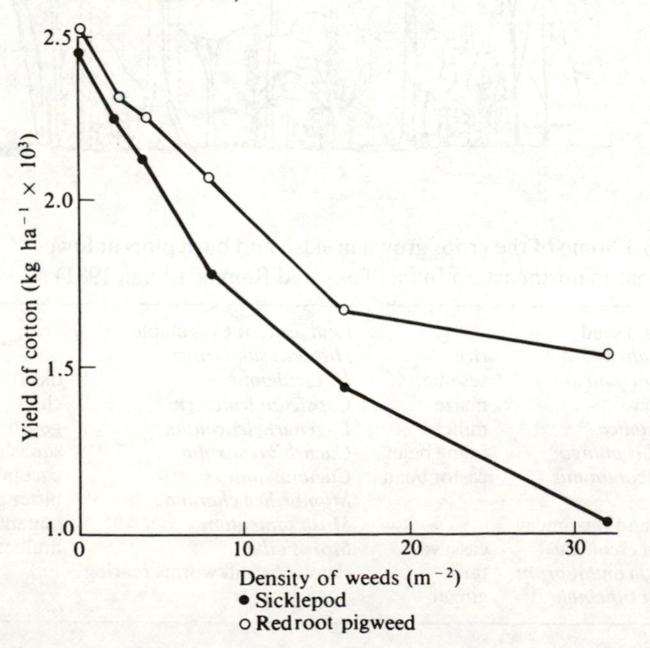

There is a problem of interpretation with an additive experiment ot this kind because the effects of total plant density (crop + weed) and of weed density on the weed (or the crop) are compounded. All plots with high weed density also have high total density. In order to ascertain the effect of the weed on itself, we must vary the *proportion* of weeds in the crop while maintaining total plant density constant and repeat the experiment at a range of densities. Experimental designs of this sort are examined later in this chapter.

Though the additive design places some limitations on the analytical information we can obtain from a competition experiment, it does have the advantage that it can simulate accurately the real situation of a crop planted at fixed density, infested with weeds. Most weedy crops are not actually two-species mixtures but multi-species mixtures containing many different weeds. Competition between species of weed as well as between weed and crop may occur, and in practice there must be the danger that removal of one weed will simply allow another weed species present to put on further growth, rather than releasing the crop from weed competition.

Haizel and Harper (1973) investigated this problem using barley (*Hordeum vulgare*), white mustard (*Sinapis alba*) and wild oats (*Avena fatua*) planted in an additive manner in 25 cm diameter pots. The experiment was rather complex in its details, but essentially it was designed to determine the effect of removing some or all of the weeds in a crop planted at 12 plants per pot and infested with 24 plants of 1 weed or 12 plants each of 2 weeds. In the agricultural situations which this experiment models, the crop is barley and the weeds are mustard and wild oats. Haizel and Harper replicated their experiment so that they could treat either barley, mustard or wild oats as the crop and the remaining species as weeds, thus generalizing the nature of the whole experiment. One-half or all the weeds were removed from pots before emergence or 3 weeks after seedling emergence, and yields of total above-ground dry weight were compared with controls from which no weeds were removed.

When barley was treated as the crop and it was grown with both weeds present, wild oats was responsible for most of the loss in barley yield. Barley yield was not depressed by the low density (12 plants) of mustard remaining when wild oats were removed from the crop before emergence but barley yields were affected by a higher mustard density (24 plants). In this experiment, as in those by Buchanan *et al.* on cotton, yield loss in the crop was not linearly related to the density of weeds. When mustard was removed from a barley crop also containing wild oats, the wild oats increased its growth and almost no improvement in barley yield occurred. Where mustard or wild oats was the only weed present in a barley crop, each responded differently when half their population was removed from pots. Even when removal was late (3 weeks after emergence), the remaining wild oat plants increased their growth to compensate for the plants removed and barley yield did not improve. Barley yield was improved by removing half the mustard weed population (12 of 24 plants) but only if this was done before emergence. These experiments suggest that partial control of wild oats or control of mustard and not of wild oats (where both weeds are present) is unlikely to improve yield in barley crops.

The unusually extensive number of combinations of species and density that Haizel and Harper used make it possible to compare the effect of each species on itself and on the other species. With barley as a crop, yield was suppressed most by added barley plants, next by wild oats and least by mustard. Treating wild oats as the 'crop', suppression of yield was in the order barley > mustard > oats, and with mustard as the crop yield was suppressed in the order mustard > barley > oats. These results show that the relative effect of different species in suppressing a crop depend upon the identity of the crop. Oats have a greater effect than mustard against barley but this situation is reversed against oats or mustard.

A contrary result was obtained by Welbank (1963) in additive experiments in which twelve weed species were grown in populations of sugar-beet, kale and wheat. Weeds which heavily suppressed yield in kale also heavily suppressed yield in wheat, despite the considerable morphological difference between these two crops. The relative effects of different weeds in kale and sugar-beet were also similar. In experiments where two crop species of oats were grown in mixtures with four weed species of *Avena*, the relative effects of different weeds were also found to be the same on the two crops (Trenbath and Harper 1973).

Separating competitive effects above and below ground

Weeds added to a crop, or plants of one species added to a population of another introduce competition from two sources: above-ground between shoots and below-ground between roots. The below-ground component of plant interactions is often ignored, though experiments demonstrate, as reason suggests, that these are important.

An early field experiment on competition between tree roots and the roots of the ground-layer vegetation in a Scottish pinewood was made by Watt and Fraser (1933). Whereas in glasshouse or other cultivation experiments the appropriate additive experimental design involves *adding* plants in mixed culture, in the field the appropriate experiment is to *subtract* plants from existing mixtures. In Watt and Fraser's experiments they effectively subtracted the influence of tree roots from a series of adjacent plots containing wavy hair grass (*Deschampsia flexuosa*) and wood sorrel (*Oxalis acetosella*) by digging trenches around each plot. These trenches were dug to various depths so as to cut off tree roots at progressively deeper levels to a maximum depth of 46 cm. Control plots were not trenched and the tree canopy was left intact in all treatments.

The roots of *D. flexuosa* extended to about 40 cm depth at the experimental site. When compared with plants in the control, plants of this species grew progressively better in plots trenched to progressively deeper levels. *Oxalis acetosella* is shallow rooting and at this site it had roots that penetrated no deeper than 7.5 cm. In spite of this shallow rooting, this species also benefited more in deeply trenched plots than in shallow ones. Watt and Fraser added distilled water to an untrenched plot and showed that this treatment did not produce the same effect on the herbs as release from interference from tree roots. This suggested that the competitive effects did not involve water as a limiting factor. On the grounds of other, incomplete, evidence, Watt and Fraser suspected that nitrogen might play a role in competition between roots of Scots pine and herbs.

Glasshouse experiments on root competition between herbs have to be designed to ensure that the effects of shoot interference are not confused with the effects of roots. Somehow the roots of plants have to be added to a pot without also adding the shoots. A simple design which permits this is shown in Fig. 8.5(a).

This additive design was used by Groves and Williams (1975) to investigate competition between subterranean clover (*Trifolium subterraneum*) and skeleton weed (*Chondrilla juncea*). The latter is a major weed of cereal crops in Southeast Australia. Part of the reason for its

Fig. 8.5 A competition experiment with subterranean clover and skeleton weed which separates the effects of root and shoot competition. (a) The four competition treatments for skeleton weed used in the experimental design. The treatments were replicated with and without a rust infection. (b) The experimental results, expressed as the dry weight of skeleton weed produced from a treatment at the final harvest as a percentage of the no-competition, rust-free treatment. (Redrawn from Groves and Williams 1975)

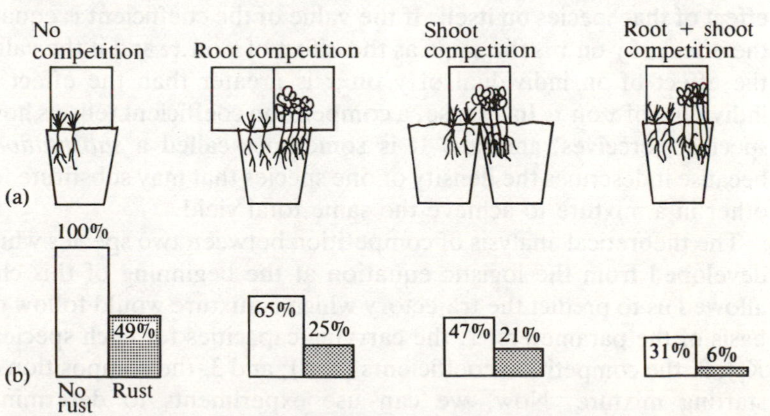

success as a weed is due to its ability to persist through the fallow period between cereal crops when fields are sown as pasture. When subterranean clover is sown as a main component of this pasture, a 60 per cent reduction in the abundance of skeleton weed can be achieved in 4 years. One of the agents used to control the weed is a strain of the rust fungus *Puccinia chondrilliana*. The interference experiment was replicated with a set of treatments containing rust-infected plants of *Chondrilla*.

The results of the experiment at the final harvest expressed in terms of relative plant dry weight of *Chondrilla* are shown in Fig. 8.5(b). *Trifolium subterraneum* suffered no significant effects of interference in any treatment and is not shown in Fig. 8.5(b). For uninfected plants, shoot interference alone had a greater effect on *Chondrilla* dry weight than root interference alone. The effect of both types of interference acting together was about what might be expected if the effects of root interference and shoot interference are multiplicative (65% root × 47% shoot = 30.6% root × shoot). Rust infection depressed dry weight to about half the control value in the no-interference treatment and by the same proportion in the shoot-interference treatment and in the root-interference treatment. Rust infection had its most severe effect in the treatment where both root and shoot interference occurred together. In this treatment the 6 per cent relative yield of *Chondrilla* was very near what is to be expected from the multiplicative effects of root and shoot interference (in rust-infected plants) (25% × 21% = 5.3%).

Additive series experiments

The competition coefficients we used in equations [8.3] and [8.4] describe the effect on a plant of an individual of the competing species, relative to the effect of another individual of its own species. If the value of a in [8.3] is <1, an individual of species y has a competitive effect on species x that is less than the effect of one individual of x. For example if $a = 0.5$, individuals of species y have an effect on species x equivalent to half the effect of that species on itself. If the value of the coefficient is equal to 1, the effect of y on x is the same as the effect of x on x, and if the value >1 the effect of an individual of y on x is greater than the effect of an individual of x on x. In a sense, a competition coefficient tells us how one species 'perceives' another. It is sometimes called a *substitution rate* because it describes the density of one species that may substitute for the other in a mixture to achieve the same total yield.

The theoretical analysis of competition between two species which we developed from the logistic equation at the beginning of this chapter allowed us to predict the trajectory which a mixture would follow on the basis of the parameters: 1. the carrying capacities for each species (K_x, K_y); 2. the competition coefficients (a, b); and 3. the composition of the starting mixture. Now, we can use experiments to determine the trajectories and then work the calculation backwards to find the competition coefficients and carrying capacities which describe an interaction.

By setting up mixtures at a range of densities and proportions, an additive series experiment (Fig. 8.1(c)) allows us to draw the trajectories on a joint abundance diagram (e.g. Fig. 8.2) for a particular pair of species. Two examples are shown in Figure 8.6. There is an equilibrium combination of *Linum grandiflorum* and *Salvia splendens* (Fig. 8.6(a)) at densities of about 10:8 in the mixtures studied by Antonovics and Fowler (1985) but this is unstable. Compare Fig. 8.6(a) with Fig. 8.2(d). Both show unstable equilibria but the isoclines for the experiment with *Linum* and *Salvia* are not straight. This implies that the logistic model is not an accurate description in this case because the competition coefficients change with density. Alternative models are discussed by Firbank and Watkinson (1985), and Law and Watkinson (1986). The experiment with the two grasses *Phleum arenarium* and *Vulpia fasciculata* (Law and Watkinson 1986) also revealed density-dependent competition coefficients. Competition between these species tended to produce a monoculture of *Vulpia* (compare Fig. 8.6(b) with Fig. 8.2(c)). Despite this result, these sand dune species do occur together in mixtures in nature. Some of the factors which greenhouse experiments omit and which can determine whether competing species coexist in nature are discussed in the next chapter.

In another additive series experiment with *Poa annua* and *Stellaria media* Connolly *et al.* (1986) found that the value of the competition

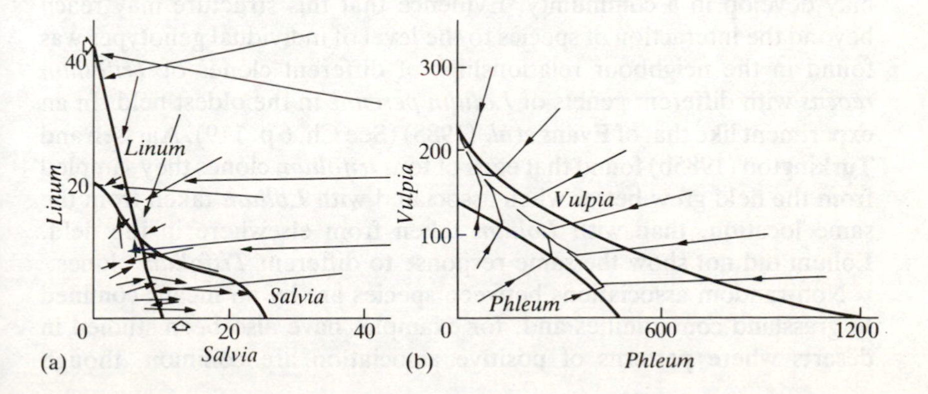

Fig. 8.6 (a) A joint-abundance diagram with trajectories showing the outcome of competition between *Salvia splendens* and *Linum grandiflorum* in different starting mixtures. Isoclines have been added to the diagram given by Fowler and Antonovics (1985). (b) A joint-abundance diagram showing the isoclines derived for an additive series experiment with *Phleum arenarium* and *Vulpia fasciculata*. The trajectories shown have been calculated from a competition model fitted to the experimental data. (Law and Watkinson 1986)

coefficients describing the interaction between the species changed with time of harvest and nutrient status of the soil as well as with density. This does not seem to be an unusual situation and implies that species cannot be attributed some fixed quality of 'competitiveness'.

Positive and negative association between species in the field

Laboratory experiments and competition theory suggest that some species can coexist side by side, while interactions between others will lead to one ousting the other. Which species is ousted will, of course, depend upon many circumstances including the original ratio of the two species present (Fig. 8.2(d)). Within a habitat we may therefore expect to find some species occurring together more often than can be predicted from chance (positive association) and others which occur together less often than predicted (negative association). This is exactly what is found in plant communities where totally random spatial association between species appears to be very rare. However, besides competition, other processes such as dispersal (Ch. 2) and clonal growth (Ch. 6) contribute to these patterns of association.

Using a plotless sampling method in an old pasture in North Wales, Turkington *et al.* (1985) found many positive and some significant negative associations between species. Using a different sampling method Stowe and Wade (1979) also found significant associations among species in a 6-year abandoned old field in Illinois. Departures from random are to be expected in the spatial distribution of species in recently established vegetation such as old fields because initial colonization is likely to be patchy. Aarssen and Turkington (1985a) compared the number of significant negative and positive associations between species in pastures 2 years, 21 years, and 40 years old, in British Columbia, Canada. All three fields contained many non-random associations between species. Most of these associations changed from year to year but some stable ones occurred in the older fields (Fig. 8.7). This suggests that, with sufficient time, a spatial structure that is the outcome of neighbour–neighbour interactions may develop in a community. Evidence that this structure may reach beyond the interaction of species to the level of individual genotypes was found in the neighbour relationships of different clones of *Trifolium repens* with different genets of *Lolium perenne* in the oldest field. In an experiment like that of Evans *et al.* (1985) (See Ch. 6 p. 119), Aarssen and Turkington (1985b) found that each of four *trifolium* clones they sampled from the field grew better when associated with *Lolium* taken from the same location, than with *Lolium* taken from elsewhere in the field. Lolium did not show the same response to different *Trifolium* clones.

Non-random associations between species are by no means confined to grassland communities and, for example, have also been studied in deserts where patterns of positive association are common, though

Fig. 8.7 Stable paired associations (———), positive; (—//—), negative detected (a) in the 21-yr-old pasture, and (b) in the 40-yr-old pasture in British Columbia. (From Aarssen and Turkington 1985a)

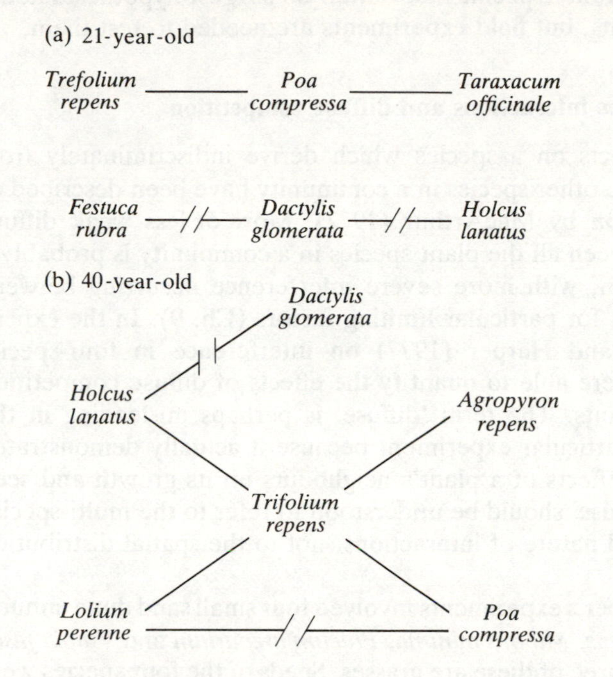

patterns of negative association are frequently interpreted as the result of competitive interactions between species (e.g. Yeaton *et al.* 1985; Cody 1986). It is actually impossible to know whether competition or some other causal mechanism lies behind a particular pattern of interspecific association just by describing or quantifying that pattern (Harper 1982). For example patterns of positive and negative association might be explained in two quite different ways.

The first possibility is that the environment is patchy and different species occupy different patches according to the pH, moisture content or nutrient status of the soil. Species with the same 'preference' for particular environmental conditions would then occupy the same patches and this would show up in the analysis as positive association. Species occupying different patches would also show up as negatively associated but neither type of association could be properly attributed to species interaction, since plant distribution would really only reflect the plants' responses to the physical environment and not to the presence or absence of other species. The second possibility is that the patchiness of the physical environment is of secondary importance and that positive and negative association between species are the results of varying levels of competition between *particular species* pairs. If this is the case one would expect to find species with affinities for the *same* type of environmental conditions competing most strongly with each other and consequently rarely occurring in contact.

Descriptions of interspecific association do suggest hypotheses about causal mechanisms, but field experiments are needed to test them.

Multi-species interactions and diffuse competition

Competitive effects on a species which derive indiscriminately from all or many of the other species in a community have been described as *diffuse competition* by MacArthur (1972). More or less weak diffuse competition between all the plant species in a community is probably a common situation, with more severe interference occurring between particular species for particular limiting factors (Ch. 9). In the experiment by Mack and Harper (1977) on interference in four-species mixtures, they were able to quantify the effects of diffuse competition on individual plants. The term 'diffuse' is perhaps misleading in the context of this particular experiment because it actually demonstrated the highly local effects of a plant's neighbours on its growth and seed production. 'Diffuse' should be understood to refer to the multi-species and multi-faceted nature of interactions, not to the spatial distribution of their effects.

Mack and Harper's experiments involved four small sand dune annuals *Cerastium atrovirens, Mibora minima, Phleum arenarium* and *Vulpia fasciculata*. The last three of these are grasses. Seeds of the four species were sown in sand in flats. Shortly after germination, seedlings were thinned to give a random distribution of plants with the four species present in equal proportions. Plants were mapped and at the end of the growing season they were harvested. The neighbourhood relationships of each plant were measured by three parameters: 1. The size of neighbours; 2. the distance separating plants from their neighbours; and 3. the distribution pattern of neighbours, determined by their location in the four quadrants of a circle around the plant. These factors were found to account for up to 69 per cent of the variance in individual plant weight. Mortality was low (5%) in this experiment, but fecundity was related to plant weight, and hence to neighbour effects, in *Vulpia*, *Phleum* and *Mibora*.

There is an unfortunate discrepancy between the type of competition experiments most often performed under cultivated garden or greenhouse conditions and conditions in the field. Two-species interference experiments or pairwise combinations among a set of species are the simplest design for cultivation experiments. But two-species experiments are often totally impractical in the field. Field experiments unavoidably involve many species. As a rule, the simplest experiments are also the best because their results are easily interpreted. Ideally then, the simplest field experiments on interspecific interference involve removing individual species from a community and monitoring the response of the remaining species.

A number of experiments of this kind have been performed. Firstly they demonstrate that the behaviour of a plant in a mixture of two species may be quite different from its behaviour in more diverse mixtures. For instance, when plantain (*Plantago lanceolata*) was removed from field plots in a grassland community in North Carolina, USA, the abundance of winter annuals increased. However, this *only* happened if sheep's sorrel (*Rumex acetosella*) was absent from the experimental plot. Where *Rumex* was present and *Plantago* was removed, *Rumex* and not winter annuals benefited from the removal (Fowler 1981). Hence in the field situation the relationship between specific pairs of species is contingent upon the presence or absence of other species. This can lead to the apparently paradoxical result that a species may decrease in abundance when another species is removed because of the effect this removal has on a third species.

Such an effect was found in the study of a community of desert annuals by Davidson *et al.* (1985) (Ch. 2, p. 29) and was also observed by del Moral (1983) when he removed *Carex spectabilis* from plots where it was dominant in a subalpine meadow community in Washington State. The grass *Festuca idahoensis* increased and produced a significant decrease in four other species. This does imply that the *Carex–Festuca* interaction was a specific and not a diffuse one.

Effects of this kind plainly depend upon which species happen to occupy a particular experimental plot. The pattern of plant distribution is important. Fowler (1981) found that up to 67 per cent of the variance in the response of a species to the removal of another from her plots was due to differences in plant distribution between plots.

On the whole, field experiments involving species removal demonstrate very few specific interactions between species that cannot be accounted for by the disposition of individual plants before removal was carried out. In the short term, the species to respond most strongly to the local removal of another may be whichever one happens to be nearest the gap which has been created. In the longer term, gaps may be colonized by plants regenerating from seed, and on this time-scale the size of gap may well determine which species appears in it.

Removal experiments in old field communities conducted by Pinder (1975), Allen and Forman (1976), Abul-Fatih and Bazzaz (1979), Hils and Vankat (1982) and experiments in grassland by Fowler (1981) all suggest that specific competitive relationships between particular species are few. For instance, Pinder found that all remaining species increased their net production by about three times when clumps of the dominant grasses in the community were removed.

In removal experiments at five different sites on a marsh on the coast of North Carolina, Silander and Antonovics (1982) found some specific responses to the removal of particular species, but these responses varied between sites. Differences in the spatial pattern and abundance of species between sites before removal probably determined the outcome. In this and the other studies, the few specific interactions between species are often not symmetrical. Goldberg and Werner (1983) argue that interspecific competitive interactions in the field are usually size-specific rather than species-specific because the relative size of a plant, and whether it or its competitors are seedlings, juveniles or larger determines the outcome of competition. This is borne out by del Moral (1983) who found that adult transplants survived better than seedling transplants and that survival of both depended upon the productivity of different sites, and hence the size of competing plants, within his subalpine meadow community. Grace (1985) found that the outcome of competition between two cattail *Typha* species was different in mixtures of juveniles raised from seed and in mixtures of adults. The importance of size asymmetry in general remains to be seen but it has important implications for the coexistence of species, which we will deal with in the next chapter.

All of the removal studies have been short-term ones and we do not know what effects the selective removal of individual species over many years would reveal. It is possible that they might be different. However, forest communities in the eastern United States where chestnut was a dominant tree changed little when this species was eliminated by chestnut blight in the first half of this century. Gaps left by chestnut were mostly occupied by bordering trees (Shugart and West 1977). Chestnut itself was migrating slowly northwards following the retreat of the last glaciation at the time it was wiped out (Davis 1981). Plant invasions and epidemic diseases occur on a far grander scale than any ecologists' experiments and the importance of competition in communities, relative to these other phenomena, must be judged in this larger perspective.

Herbivory and competition between plants

Herbivores rarely kill growing plants directly (Ch. 5) but they can alter the outcome of competition between different species and consequently influence the composition of vegetation greatly. The beetle *Gastrophysa viridula* grazes upon two species of dock, *Rumex obtusifolius* and *R. crispus*. The docks frequently occur together in Britain where Bentley and Whittaker (1979) found that grazing by beetles had a more severe effect upon the plants when growing in competition with each other than when grown alone. Though *Gastrophysa* prefers *R. obtusifolius*, the interactive effect of grazing and competition upon *R. crispus* was such that this species suffered most.

Insect herbivory may accelerate or impede changes in vegetation composition. Pastures sown with perennial ryegrass *Lolium perenne* are usually invaded by other species, particularly dicots. Clements and Henderson (1979) found that this change occurred more slowly in plots from which insects were eliminated with pesticide. Herbivores, in particular the stem-boring larvae of the fly *Oscinella frit* which attack grasses, reduce grass growth and its ability to resist invasion. Oldfields in northeastern North America are invaded by goldenrods (*Solidago* spp.) which tend to displace the grasses that occur in earlier stages of succession. McBrien *et al.* (1983) observed an outbreak of three chrysomelid beetle species (*Trirhabda* spp.) which severely defoliated *Solidago canadensis* and allowed an increase in grass cover. Clones of *Solidago canadensis* vary significantly in their susceptibility to insect attack (D. Maddox and R. Root 1987) so that herbivory may change the genotypic composition of populations as well as the species composition of communities.

Vertebrate herbivores play a crucial role in determining the relative abundance of species in plant communities. Examples of this role are legion (Harper 1977; Crawley 1983) but none is more dramatic than the decline in species numbers observed by Watt (1974) over a period of nearly 40 years following the exclusion of rabbits from a plot of chalk grassland in East Anglia (Fig. 8.8).

Interspecific interactions that affect plant performance (e.g. yield) clearly occur wherever neighbours physically encroach upon each other, but the significance of this for the population dynamics of the species involved depends upon many factors such as herbivory, population structure and the spatial distribution of plants. The next chapter takes the larger, more theoretical perspective needed to explore the significance of competition in the field.

Fig. 8.8 The number of original species surviving in a plot of chalk grassland following the exclusion of rabbits in 1936 (•). New species that became established in the plot (○). Species temporarily invading mole hills (×). From Watt (1974).

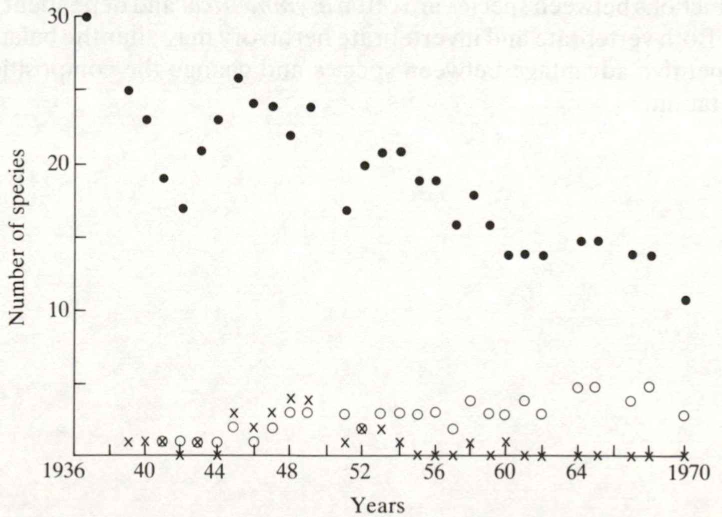

Summary

Definitions of *competition* differ, tending to stress either the *mechanism* or the *outcome* of the interaction. The latter kind of definition is more appropriate to population ecology but mechanisms and the fact that competition occurs between *neighbours* should not be ignored. Changes in numbers of two competing species with time can be represented by a *trajectory* on a *joint-abundance diagram*. The *Lotka–Volterra equations* predict that two competing species with similar carrying capacities will stably coexist if, at high density, each inhibits its own population growth more than that of its competitor. Three designs of competition experiment between plants are commonly used. In *replacement series experiments* the proportions of two species are varied at a fixed total density. In *additive experiments* the density of one species is fixed and that of the other is varied. The *additive series* is equivalent to a replacement series repeated at a range of densities and is the only design to explore the full range of density and proportion in mixtures.

Relative Yield Total (RYT) is calculated from replacement series experiments but it can only be used to measure performance in mixtures under restrictive conditions. The *Land Equivalent Ratio* (LER) is a similar measure to RYT but is more flexible. Judged by LERs, many intercrops yield better than comparable monocultures. Competition between species in additive series experiments is measured by *competition coefficients* which describe how each species 'perceives' the other. Competition coefficients may vary with density.

Associations between species in the field are rarely random, though competition is only one of many factors that may be responsible for this. *Removal experiments* can be used to investigate interspecific interactions in the field. They usually demonstrate that competition is local and not species specific (i.e. it is '*diffuse*'), though exceptions to this do occur. Interactions between species are often *asymmetrical* and dependent upon size. Both vertebrate and invertebrate herbivory may alter the balance of competitive advantage between species and change the composition of vegetation.

9
Coexistence and the niche

The richest plant communities contain a plethora of species: 1316 native plants have been recorded in 15 square kilometres of tropical forest on Barro Colorado Island (BCI), Panama (Table 9.1); the most diverse tropical forests contain 180 or more species of trees in a single hectare. This tropical species richness is often contrasted with a temperate paucity; the entire native flora of the British Isles contains only about 1600 plant species in an area 314 375 km^2, but the temperate zone does have species-rich herbaceous vegetation. Calcareous grasslands in Britain may contain up to 30 species packed into 1/8th square metre, pine–wiregrass savanna in North Carolina may have up to 42 species in the same area (Walker and Peet 1983). The major differences in diversity between temperate and tropical zones have an historical and climatic basis (e.g. Silvertown 1985). Examples of locally high species density seem to occur wherever history has provided a source of many taxa, and climate has permitted their survival: trees in the tropics, herbs in the temperate zone. Is there really no intrinsic limit to the number of plant species which may coexist in a habitat? According to Lotka–Volterra theory of competition there ought to be a limit, but there is a staggering discrepancy between the way this model of competition, based upon the logistic equation (Ch. 8, p. 160), would have plants behave and the real world in which species-rich vegetation is common. In this chapter we will explore the apparent contradiction between the results of competition theory and real vegetation and look at some of the resolutions to the problem that have been proposed.

Solutions to the question fall into two groups: 1. *equilibrium models* which assume that populations remain within defined limits of abundance; and 2. *non-equilibrium* models which invoke local fluctuations in population size to explain coexistence.

Equilibrium models: The competitive exclusion principle and the niche

In Chapter 8 we saw that, according to Lotka–Volterra theory, the key to the coexistence of two competing species in a stable equilibrium is that each should inhibit its own population growth more than that of the other. One way for this to occur is that the population growth of each

Table 9.1. Life forms of plant species at Barro Colorado Island, Panama

	No. of species	%
Cryptogams		
Epiphytes	41	3.1
Hemi-epiphytes	1	0.1
Aquatics	6	0.5
Vines	4	0.3
Other terrestrials	47	3.6
Tree ferns	5	0.4
Total cryptogams	104	7.9
Phanerogams		
Trees >10 m tall	211	16.0
Trees <10 m tall	154	11.7
Shrubs 2(–3) m tall	93	7.1
Epiphytic or hemi-epiphytic trees and shrubs	16	1.2
Parasitic shrubs	7	0.5
Total arborescent spp.	481	36.5
Lianas or woody climbers (incl. 10 climbing trees)	171	13.0
Vines	83	6.3
Epiphytic or hemi-epiphytic vines	11	0.8
Total scandent spp.	265	20.1
Epiphytic herbs	135	10.3
Aquatic herbs	54	4.1
Herbs of clearings	197	15.0
Forest herbs	75	5.7
Parasitic herbs	1	0.1
Saprophytic herbs	4	0.3
Total herbs	466	35.4
Total all native plant species	1316	100

Source: Croat (1978).

species must be limited by a different *limiting factor*. Nutrients and space are common limiting factors for plants. For example, nitrogen often limits grasses and potassium often limits legumes, which obtain nitrogen from their root nodules. This difference between legumes and grasses may partly explain the ability of white clover (*Trifolium repens*) to form stable associations with several grass species in the field (see Fig. 8.7). The genotypic differences between the distinct clones of white clover which are found with different grasses (Ch. 5, Ch. 8) imply that there is a complex relationship between coexisting individuals which may also involve other factors such as their phenology.

Fig. 9.1 A two-dimensional niche diagram showing the use of two mineral resources by a legume and a grass.

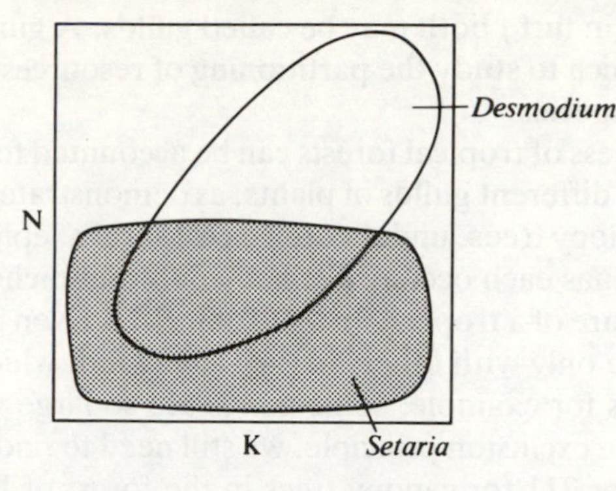

The resources a population requires to maintain or increase its size and its manner of exploiting these, when comprehensively catalogued, describe its *niche*. A useful way of visualizing a plant's niche is to imagine it as a space with two (or more) dimensions, each dimension representing the quantity of a resource. For example we might depict the relationship between a legume and a grass utilizing N and K as in Fig. 9.1. This illustrates a situation in which there is niche overlap which would lead to competition for N and K between the species when the availability of K is low. When more K is available, there is less overlap on both resource axes because a sufficient K supply allows the legume to obtain extra nitrogen from its root-nodule bacteria. The niche diagram in Fig. 9.1 is a hypothetical one, but Hall (1974) found a relationship such as this between the legume *Desmodium intortum* and the grass *Sectaria anceps* grown in experimental mixtures.

Where two species share exactly the same niche, or in other words the same factor(s) limit population growth in each, the Lotka–Volterra theory predicts that they will compete severely and one of them will be eliminated. This conclusion is known as the *competitive exclusion principle* (Hardin 1960) or Gause's principle (Gause 1934). Now we begin to see where the disagreement between theory and reality might lie. There are fewer than ten major nutrients and these are required by all plants. Do the forty-two species occurring together in 1/8th square metre of pine–wiregrass savanna occupy forty-two different niches?

Equilibrium models: Guilds and niches

Species which divide a shared limiting resource between them are described collectively as a *guild*. For example, earth mounds created by badgers in tall-grass prairie in the USA are colonized by a characteristic group of species, some of which are not found in the surrounding, undisturbed grassland (Platt 1975). Patches of bare ground created by disturbance in chalk grassland in Southern England also have their characteristic species (Grubb 1976). Since each of these sets is confined to a specific resource (gaps in turf) both may be called guilds. A guild is a convenient unit within which to study the partitioning of resources and niche separation between species.

Some of the species richness of tropical forests can be accounted for by the fact that there are many different guilds of plants, as demonstrated by the variety of life forms: canopy trees, understorey trees, shrubs, epiphytic trees and shrubs, and lianas each occupy distinctly different niches in the complex physical structure of a tropical forest (Table 9.1). Even if we assume that plants compete only with others in the same guild, which is obviously not true of lianas for example, some guilds are so large that, according to the competitive exclusion principle, we still need to find 265 niches for climbing plants or 211 for canopy trees in the forest of BCI.

A species' niche cannot be defined entirely in terms of physical environmental factors. Where a plant grows may be influenced by competition from other species in regions of niche overlap and also by predators. These factors may be incorporated into the concept of the niche if we envisage it as a large niche space incorporating all those conditions in which a population may exist when its competitors and predators are absent, containing a smaller space representing the niche of a population when its competitors and predators are present. The larger niche space is called the *fundamental niche* and the part of this which is occupied when competitors and predators are present is the *realized niche* (Hutchinson 1957).

Experiments in which competitors and predators are removed may be used to determine the extent of a plant's fundamental niche and what factors determine the limits of its realized niche. An experiment along these lines, employing two species of bedstraw (*Galium*) was performed by Tansley (1917), a full 40 years before Hutchinson's definition of the realized niche drew attention to the subject. *Galium saxatile* is a plant confined to more acidic soils in Britain whereas another species *G. sylvestre* is found in calcareous habitats. Tansley grew each species in soil from its native habitat and in soil from the other habitat in monocultures and in mixtures of the two species together. Both species grew on both types of soil when sown alone but when sown in the presence of the native species the alien species was suppressed by interference from the native and only the native species grew successfully.

It is rarely as easy to discover the factors which confine a species within its realized niche as Tansley's experiment suggests, because plants growing in the same habitat do not generally show responses to environmental variation which are as clearly defined and well differentiated as the bedstraws' response to soil pH. For instance, among the plants of rivers and streams in Britain and North America, species are differently distributed according to a large number of factors of which only a few are: the speed of water flow, substrate type, width and depth of channel, distance from the mouth of the river, turbidity of the water and pH (Haslam 1978).

Quite complex statistical procedures are often used to determine which combination of environmental variables accounts best for the observed distribution of species (Gauch 1982). In theory at least, these techniques could help us define the realized niche of a plant though, being entirely descriptive, they cannot be used to map a plant's fundamental niche (Austin 1985).

The ecologist looking for niche differences to explain coexistence must beware of a circularity in niche theory. The competitive exclusion principle states that no two species, sharing the same niche, may also share the same habitat (i.e. coexist). There is a danger that with this principle already firmly planted in his or her mind, the ecologist assumes that the principle *must* apply in a particular case and species which coexist must *therefore* have different niches. By scrutinizing coexisting species more and more closely, examining their relative rooting depth, flowering time, pollinating agents and every character of any conceivable relevance, enough differences can often be detected to 'explain' their coexistence. Measure enough characteristics of any two objects or any two species with an accurate enough scale and you are sure to find differences between them. There are statistical methods for determining the likely competition that particular levels of niche overlap will produce (Horn 1966; MacArthur and Levins 1967; May 1975), but these really only offer circumstantial evidence that competitive exclusion will or will not, should or should not take place *if* the correct characteristics have been measured. We must use other methods to judge whether niche overlap in particular resource dimensions is significant in the dynamics of plant populations. In fact the only reliable tests are experimental ones.

There is no surer way of telling whether a plant population can exist under certain conditions and whether particular factors are relevant in defining its niche than placing it in those conditions and finding out. The site of the transplants and the controls should differ in some measurable or definable way likely to represent an important limiting factor in natural populations. The fate of these transplants or *phytometers* then tells us how the habitat 'looks' from the point of view of the plant itself.

Phytometer populations often show qualitative changes or only gradual mortality in their new locations and several years may elapse before whole transplant populations actually become extinct. Measurements of the growth of phytometers may therefore be necessary to predict their long-term survival. Clymo and Reddaway (1972) studied a guild of four species of *Sphagnum* moss which occupy different niche positions in relation to microtopography in mires in the northern

Pennines of England. They measured the growth in 1 year of phytometers of each species placed in its usual microhabitat and in the habitats of the other species (Fig. 9.2). Assuming that dry matter increase of phytometers is a realistic indicator of population growth, the results suggest that species normally occupy niche positions in which they are competitively superior to other species and that interspecific interference determines the boundary of each species' realized niche. Note that a species such as *S. rubellum* may be forced to occupy a

Fig. 9.2 The growth of four *Sphagnum* species transplanted into native (hatched) and non-native parts of the habitat in which they occur. (From Clymo and Reddaway 1973)

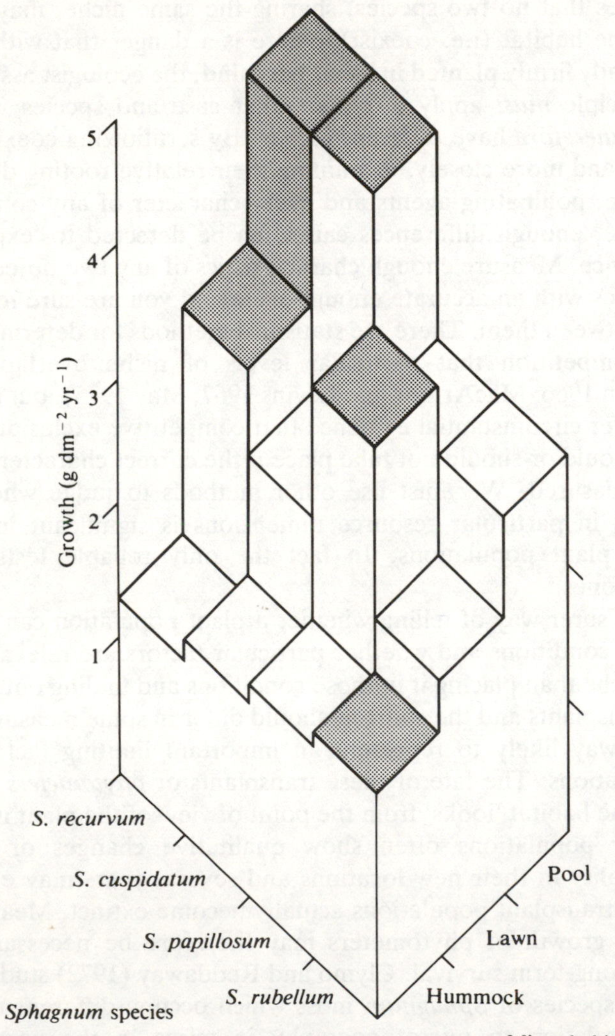

sub-optimal part of the habitat by this interference. Species are not necessarily found in those niche positions where they grow at their *individual best*, but where they grow *relatively better* than (or as well as) other species in the guild. Another example of this is found in the bluebell *Hyacinthoides non-scripta* which is commonly found in woodlands. When phytometers of this species are planted outside a wood, free from the interference of other species, they grow better than plants grown in the wood. In Britain the species is usually only found in woodlands, to which it appears to be confined by the interference of plants growing on the woodland margin (Blackman and Rutter 1950).

Oak–hickory forest in North America contains several different oak species, a number of which have local distributions which do not overlap. An effective phytometer experiment was carried out by Bourdeau (1954) to determine the factors responsible for niche differences between oak species growing at two contrasting sites in the piedmont of North Carolina, USA. *Quercus rubra* and *Q. coccinea* grew in forest with a closed canopy on a rich, damp, Georgeville-type soil whose surface was covered in deep leaf litter. *Quercus stellata* and *Q. marilandica* grew in forest with an open canopy on poor, dry Orange-type soil with only a shallow covering of leaf litter. Trees of one soil type did not occur in forests on the other soil type.

Seedlings of all four species were planted in field plots in forest on Orange soil and in forest on Georgeville soil. Phytometers of all four species survived well on the rich soil, but the survival of the native species (*Q. stellata* and *Q. marilandica*) was superior to that of the two non-native ones on poor soil. The drought tolerance of the four species, assessed in a laboratory experiment, showed seedlings of the rich-site species to be more susceptible to drought than those from the poor site. This was therefore probably the factor responsible for the poor performance of *Q. rubra* and *Q. coccinea* phytometers at the poor site. As for *Q. stellata* and *Q. marilandica*, phytometers of these poor-site species survived well in rich forest and some other factor must account for their absence from such sites. Bourdeau measured the growth of seedlings of each species in pots of Georgeville soil placed beneath the shade of a closed tree-canopy. This experiment showed that the poor-site species grew little under shade and that the rich-site species grew well. Laboratory measurements of photosynthesis in *Q. rubra* and *Q. marilandica* at a range of light intensities showed that the optimum light intensity for growth of the rich-site species was half that of the poor-site species. Complementary environmental constraints, probably reinforced by competition, therefore exclude the oaks of rich sites from poor sites and those of poor sites from rich ones.

Two generalizations emerge from the results of phytometer experiments. Firstly, plants transferred to an alien part of the habitat or to an entirely new habitat may survive for quite a long time before they die and the population 'retreats', under the impact of various factors, back within the boundaries of the realized niche. Darwin pointed out the importance of competition from other plants in this process when he observed, in *The Origin of Species* (1859), that many alien plants (exotics) may be grown quite successfully in gardens where they are freed from interference but that they cannot survive outside gardens where they must contend with native species. The second generalization that may be drawn from phytometer experiments, particularly that of Bourdeau using oak seedlings, is that populations occurring outside their realized niche are most vulnerable to extinction in the regeneration phase of the life-cycle. This is consistent with asymmetrical interference between mature plants and seedlings and the low survivorship of seedlings often observed in populations, even when growing in the native habitat (Ch. 5). Gardeners not only tend their exotic specimens by freeing them from weed interference but they often must also repeatedly sow plants in order to maintain garden populations which cannot regenerate naturally.

Equilibrium models: Regeneration niche

The vulnerability of plants in the regeneration phase and the ability of individual plants, once established, to survive despite quite severe odds against them suggests that coexistence between some species may occur because of differences in their regeneration behaviour. For example if species successfully establish only in vegetation gaps of a particular size, or only in those appearing at a particular time, differences between species may be great enough to allow seedlings to avoid interspecific competition until they are large enough to be invulnerable to it. Grubb (1977) has called this aspect of a plant's niche the 'regeneration niche'. Hobbs and Mooney (1985) found differences in regeneration niche between annual species colonizing disturbances made by gophers in a California grassland. The relative abundance of species varied with the time of year the gopher mounds were created (Fig. 9.3). Species such as the

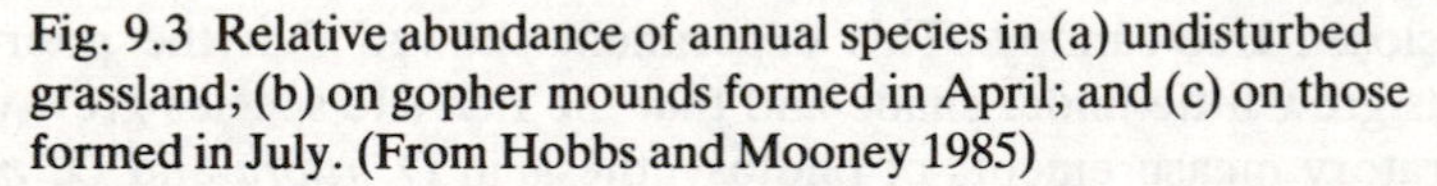

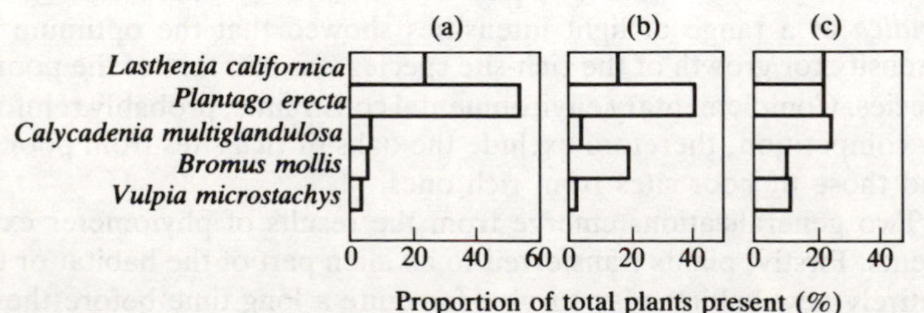

Fig. 9.3 Relative abundance of annual species in (a) undisturbed grassland; (b) on gopher mounds formed in April; and (c) on those formed in July. (From Hobbs and Mooney 1985)

Plantago spp. with different safe sites for seeds (Fig. 2.12) occupy different regeneration niches, but whether these differences are sufficient to prevent competitive exclusion when seedlings grow large enough to interfere with each other is not known.

In the guild of plants which colonize badger disturbances in tall-grass prairie, the first plants to arrive on a new mound have the highest success, virtually regardless of species. Platt and Weiss (1985) found that an average of about 70 per cent of seedlings of *Solidago rigida* arriving in the first year survived to flower, but none of those germinating on older mounds with plants already established on them did so. The first plants to arrive *preempt* the site and exclude others. When different species arrived together in the first year, competition depressed growth and seed yield, affecting species with the smallest seeds most, but not affecting survival. Species with smaller seeds may be able to avoid competition from those with larger ones by more efficient dispersal to sites which occur at low density where they can develop to an invulnerable size before other species arrive. This appears to be an example of a guild of species coexisting by partitioning a resource (mounds). Each species' niche consists of those mounds it can reach before other species that have seeds larger than itself. However, this is a highly dynamic situation of a kind in which niche separation may be unnecessary for coexistence because all the species are relatively uncommon and rarely encounter each other (Shmida and Ellner 1984).

Many plants including forest trees require gaps in vegetation in order to establish successfully (Ch. 5) and, like badger mounds, these sites are scattered. Could the coexistence of tropical forest trees be explained by differences in regeneration niche between species? Hubbell and Foster (1985, 1986) could find few differences in the gap-'preferences' of trees in the same guild in BCI and concluded that their regeneration niches were 'for all practical purposes, functionally identical.' (Hubbell and Foster 1986).

Equilibrium models: Resource ratio hypothesis

Although there are only a handful of essential mineral nutrients and all are needed by most plants, it is possible that nutritional differences between plants do play a role in niche separation in some communities. According to Lotka–Volterra competition theory, *n* competing species cannot coexist on fewer than the same number of limiting resources. How then, can forty-two species of plants in pine–wiregrass savanna coexist on fewer than ten limiting nutrients? The answer could lie in local, small-scale spatial variation in the concentration of these nutrients.

The growth rate of a plant may be limited by different nutrients under different conditions of nutrient supply. For example nitrogen may limit growth when its concentration is low but addition of this nutrient may cause an increase in growth rate to a point at which phosphorus becomes limiting. Which of two nutrients limits the growth of a plant in a particular location will depend upon the relative values of two *ratios*: 1. the ratio of concentrations at which the switch between limitation by N and limitation by P occurs, determined by the physiology of the plant; and 2. the ratio of N and P concentrations available in the soil. Species with qualitatively similar nutrient requirements (i.e. all requiring N, P, K, Mg, Ca) but responding to quantitatively different ratios may then be limited by different nutrients if the ratio of nutrients available where they are growing has a value that lies *between* the critical values at which the two species switch (Tilman 1982). When this condition is fulfilled, the difference in limiting factors allows them to coexist. If two species have the same limiting nutrient, the one with the lower requirement will displace the other. Only two species may coexist in any patch, one with a critical ratio above and one with a critical ratio below the ratio of nutrient availability. However, if the concentration of the two nutrients is locally very patchy, the supply ratios will vary from patch to patch and different pairs of species may occur in each one. An entire habitat may then support many species overall (Fig. 9.4).

Topographical patchiness has the same tendency to allow species to carve out niches for themselves and to promote coexistence, as we observed with *Sphagnum* spp. (Fig. 9.2). The unique feature of the resource ratio hypothesis, due to Tilman (1982), is that it is potentially an explanation of coexistence in species-rich habitats such as chalk grassland where there is very little topographical patchiness. The evidence in favour of the hypothesis as an explanation of coexistence in these kinds of plant communities is entirely circumstantial at the moment. The hypothesis predicts that species-rich vegetation should occur on relatively infertile soil because nutrient ratios become less patchy as nutrients are added to a site. This is born out by experiments in grassland (e.g. Silvertown 1980c) and the distribution of species-rich communities which occur on infertile soils in many parts of the world.

Equilibrium models: Aggregation hypothesis

The Lotka–Volterra model of competition ignores the spatial distribution of individuals and deals with interactions between species as though they were intimately mixed. In fact most plant populations are aggregated at some spatial scale and clumps of conspecific individuals are common, if not the rule. Bearing in mind that plants can only compete with their neighbours, it is obvious that plants will compete more and more with others of the same species and less and less with individuals of different species, the more aggregated is a population. The effect of aggregation is therefore to reduce *inter*specific competition and to increase *intra*specific competition, and consequently to increase the likelihood that the condi-

Fig. 9.4 A 'field' containing four patches, each with a different ratio of N:P in the soil. Species A, B and C have different critical ratios at which they switch from limitation by N to limitation by P, as shown on the scale. No more than two species coexist in any one patch but all three species are present in the field as a whole.

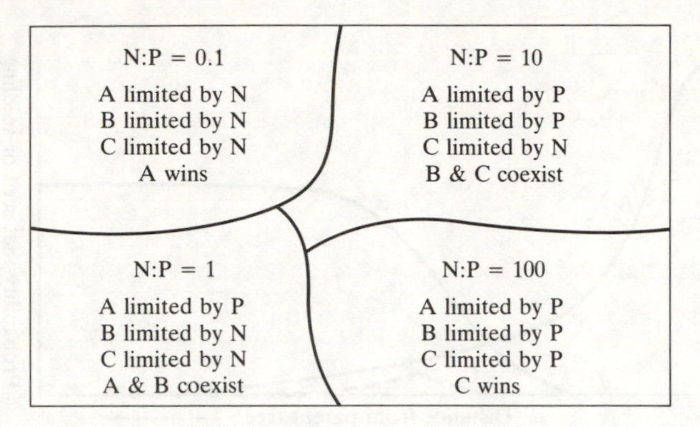

Critical ratios of N:P for three species, A, B & C:

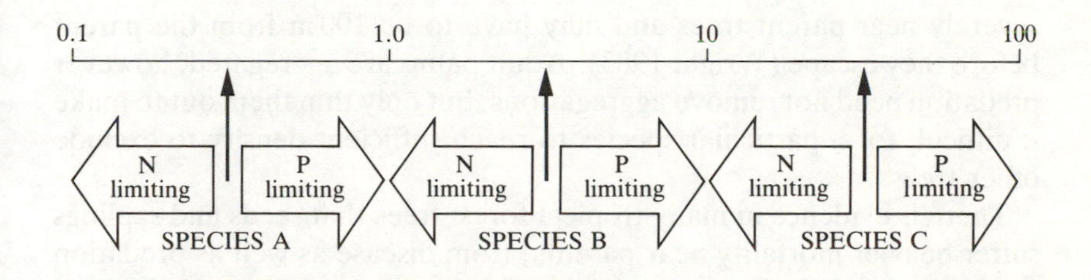

tions for coexistence we derived from the Lotka–Volterra model will be satisfied. This simple hypothesis might be one of the most general reasons for coexistence because, even ignoring other causes of aggregation, limited seed dispersal that produces clumped distributions of seeds is so much the rule (Shmida and Ellner 1984).

Equilibrium models: Density- and distance-dependent predation and disease

Whilst aggregation may promote coexistence so, paradoxically, can agents which tend to destroy aggregations of seeds and seedlings. Most seeds fall near the parent plant, but parents are also the headquarters of specialist herbivores and of seed-eating animals. Janzen (1970) and Connell (1971) suggest that these take most of the seeds near parent tropical trees, the only ones to escape being the few which have been carried a long way off by dispersal agents (Fig. 9.5) and which occur in gaps or under other species. This would facilitate coexistence between species.

Fig. 9.5 A model showing the probability of a seed (or seedling) escaping predation as a function of distance from the parent tree. A The distribution of seeds; B the probability of a seed escaping predation as a function from the parent; C the product of curves (A) and (B) is the distribution of surviving seeds. (After Janzen 1970)

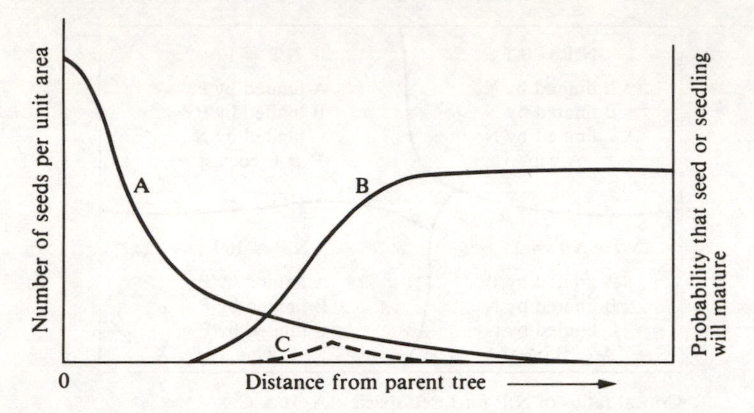

Seeds of *Scheelea* palm at BCI are attacked by bruchid beetles more severely near parent trees and may have to be 100 m from the parent before they escape (Wright 1983). Adult palms are aggregated, however predation need not remove aggregations, but only thin them out to make it difficult for a particular species to reach sufficient density to exclude other trees.

There is evidence in many tropical forest trees that seeds and saplings suffer heavier mortality near parents, from disease as well as predation (Augspurger 1984; Clark and Clark 1984; Becker and Wong 1985). Two factors are potentially important: distance to the parent and the local density of offspring. Obviously these two factors tend to be correlated, but Clark and Clark (1984) were able to assess them separately for *Dipteryx panamensis*, a canopy rainforest tree in Costa Rica, and found seedling density rather than distance-to-parent to be the important variable. At BCI, Hubbell and Foster (1986) found evidence that juvenile density was inversely related to adult density in twenty-one of the forty-eight commonest trees. The effects were not strong ones, but the test for density dependence was a correlative one prone to the weakness discussed in Chapter 4. Only a few species in Australian tropical rainforests studied by Connell *et al.* (1984) showed distance or density effects of the kind likely to promote coexistence.

By contrast, Burdon and Chilvers (1974) believe that species-specific pathogenic fungi may contribute to coexistence in species-rich Australian eucalypt forest. Augspurger (1983b, 1984) found that pathogens were a major cause of mortality to seedlings of six of nine forest trees she studied at BCI. In two cases fungi were responsible for higher mortality near parents but in all six attacked species, disease was much less important in light gaps than under the canopy, even though densities were quite high in gaps. Recruitment in gaps is therefore likely to generate a clumped distribution of adults in these species.

Non-equilibrium models

All the hypotheses we have discussed attempt to explain coexistence in equilibrium conditions, that is without assuming that population density fluctuates. In fact all populations fluctuate at some scale, but this phenomenon does not enter into explanations of coexistence based upon niche separation, aggregation or density dependence. We now look at a group of models which do not assume equilibrium and in which coexistence does depend upon fluctuations.

Non-equilibrium models: Disturbance

Forests and most other types of vegetation are prone to periodic, though unpredictable, disturbance which kills adults and creates opportunities for recruitment (Pickett and White 1985). Huston (1979) examined the effect of disturbance upon the outcome of competition between two species sharing one resource in a model based upon the Lotka–Volterra equations. Without disturbance the model led to the extinction of one species (Fig. 9.6(a)), but a density-independent population reduction (e.g. removal of 50 per cent of each population) led to prolonged coexistence between them (Fig. 9.6(b)), though one did eventually go extinct. A model with six competing species, sharing the same limiting resource,

Fig. 9.6 The outcome of competition between two species based upon a Lotka–Volterra model: (a) without disturbance; (b) when there is a disturbance producing a periodic, density-independent population reduction. (After Huston 1979)

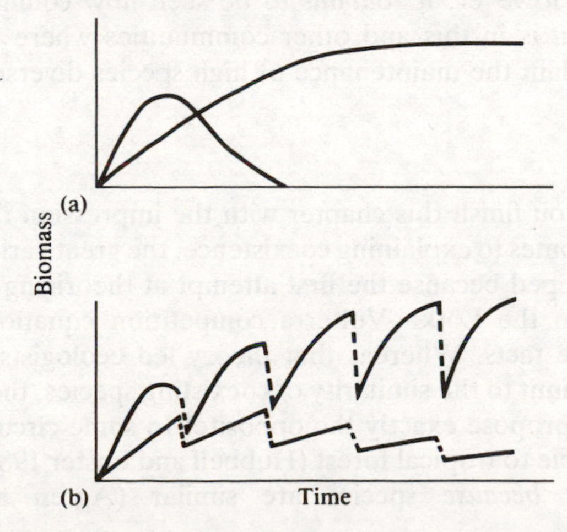

showed that these could coexist for a long period provided that: 1. all population growth rates were low; and 2. disturbance was not too frequent.

This model fits well with observed trends of diversity in forests where the highest species richness seems to occur at intermediate frequencies and intensities of disturbance (Connell 1978). The highest diversities of plants in pine–wiregrass savanna occur in annually burned sites (Walker and Peet 1983) and local disturbances by frost-heave increases diversity in subalpine and alpine communities (del Moral 1983; Fox 1981).

Non-equilibrium models: The storage effect

Fluctuations in recruitment can in themselves promote coexistence if 'good years' and 'bad years' for different species occur out of synchrony. In a good recruitment year a species can establish a cohort of juveniles in the gaps currently available and once grown beyond small size these individuals may be invulnerable to competition with juveniles of other species which appear some time later. This is known as the *storage effect* because recruits born in good years are 'stored' through the bad years when competition from other species can prevent recruitment (Warner and Chesson 1985).

In addition to fluctuating recruitment, for the storage effect to work and to promote coexistence plants must have: 1. overlapping generations (otherwise recruitment has to happen every generation and storage is not possible); and 2. competition between adults must be weak. Many perennial herbs, annuals if they have a seed pool and trees (especially those with oskars – Chs. 2 and 5) appear to fit these criteria. Comins and Noble (1985) attribute coexistence among Australian eucalypts to the storage effect and Grubb (1986) believes it may be responsible for the coexistence of chalk grassland perennials which do have unsynchronized years of good seed production. However, it remains to be seen how common asynchronous *recruitment* is in this and other communities where the storage effect might explain the maintenance of high species diversity.

Conclusion

You need not worry if you finish this chapter with the impression that 'anything goes' when it comes to explaining coexistence: the great variety of hypotheses has developed because the first attempt at theorizing on this subject, based upon the Lotka–Volterra competition equations, patently didn't fit all the facts. Whereas that theory led ecologists to believe that there was a limit to the similarity of coexisting species, there are now theories which propose exactly the opposite! In some circumstances, perhaps applicable to tropical forest (Hubbell and Foster 1986), coexistence may occur *because* species are similar (Agren and

Fagerstrom 1984). Not all these theories (and their variants which you have been spared) will stand the test of time, but they should all be tested. It is quite likely that none of the theories is universally responsible for the patterns of diversity we see in nature. In this author's view pluralism is essential in ecology. The diversity of nature and the increasing diversity of ecological theory are an exciting challenge.

Summary

High plant species diversity is at odds with the *competitive exclusion principle*. Two kinds of theory exist to explain the discrepancy: *equilibrium models* and *non-equilibrium models*.

Equilibrium models invoke *niche* differences between species, particularly those species in the same *guild* which compete for and share a resource. The concept of the niche can be resolved into the *fundamental niche* and the *realized niche*. *Phytometer* experiments can be used to map out a species' niche and to determine the effects of interference from other species. *Regeneration niche* refers to the characteristics of reproduction and establishment, which tend to differ between species. The *resource ratio hypothesis* may explain the coexistence of several species which compete for two resources if the resources are patchily distributed. Topographical patchiness may operate with similar effect. *Aggregation* may promote coexistence by increasing *intra*specific and reducing *inter*specific competition. *Density-* or *distance-dependent* predation and disease may facilitate coexistence between trees in tropical forests.

Disturbance operating in a density-independent fashion may facilitate coexistence by reducing the dominant competitors. Fluctuations in recruitment allow populations to survive bad years by strong recruitment in good ones. This mechanism, known as the *storage effect*, may prevent competitive exclusion in species where seedlings are vulnerable to the effects of competition but older plants are not.

Bibliography

Aarssen, L.W. and Turkington, R. (1985a) Vegetation dynamics and neighbour associations in pasture–community evolution, *J. Ecol.*, **73**, 585–603.

Aarssen, L.W. and Turkington, R. (1985b) Biotic specialization between neighbouring genotypes in *Lolium perenne and Trifolium repens* from a permanent pasture, *J. Ecol.*, **73**, 605–14.

Abrahamson, W.G. (1980) Demography and vegetative reproduction, Ch. 5, pp. 89–106, in Solbrig, O.T. (ed.) *Demography and evolution in plant populations*. Blackwell, Oxford; University of California Press, California.

Abrahamson, W.G. and Gadgil, M. (1973) Growth form and reproductive effort in goldenrods (*Solidago*, compositae). *Am. Nat.*, **107**, 651–61.

Abul-Fatih, H.A. and Bazzaz, F.A. (1979) The biology of *Ambrosia trifida* L. I. Influence of species removals on the organization of the plant community, *New Phytol.*, **83**, 813–16.

Agren, G.I. and Fagerstrom, T. (1984) Limiting dissimilarity in plants: randomness prevents exclusion of species with similar competitive abilities, *Oikos*, **43**, 369–75.

Alexandre, D.Y. (1978) Le rôle disseminateur des éléphants en forêt de Tai, Côte-d'Ivoire, *Terre Vie*, **32**, 47–72.

Allen, E.B. and Forman, R.T.T. (1976) Plant species removals and old-field community structure and stability, *Ecology*, **57**, 1233–43.

Alpert, P. Newell, E.A. Chu, C. Glyphis, J. Gulmon, S.L. Hollinger, D.Y. Johnson, N.D. Mooney, H.A. and Puttick, G. (1985) Allocation to reproduction in the chaparral shrub, *Diplacus aurantiacus, Oecologia*, **66**, 309–16.

Anderson, R.C. and Loucks, O.L. (1973) Aspects of the biology of *Trientalis borealis* Raf., *Ecology*, **54**, 798–808.

Angevine, M.W. (1983) Variations in the demography of natural populations of the wild strawberries *Fragaria vesca* and *F. virginiana*, *J. Ecol.*, **71**, 959–74.

Antonovics, J. and Ellstrand, N.C. (1984) Experimental studies of the evolutionary significance of sexual reproduction. I. A test of the frequency-dependent selection hypothesis, *Evolution*, **38**, 103–115.

Antonovics, J. and Fowler, N.L. (1985) Analysis of frequency and density effects on growth in mixtures of *Salvia splendens* and *Linum grandiflorum* using hexagonal fan designs, *J. Ecol.*, **73**, 219–34.

Antonovics, J. and Levin, D.A. (1980) The ecological and genetic consequences of density-dependent regulation in plants, *Ann. Rev. Ecol. Syst.*, **11**, 411–52.

Arthur, A.E. Gale, J.S. and Lawrence, K.J. (1973) Variation in wild populations of *Papaver dubium*: VII. Germination time, *Heredity*, **30**, 189–97.

Ashmun, J.W. Thomas, R.J. and Pitelka, L.F. (1982) Translocation of photoassimilates between sister ramets in two rhizomatous forest herbs, *Ann. Bot.*, **49**, 402–15.

Ashmun, J.W. and Pitelka, L.F. (1985) Population biology of *Clintonia borealis*. II. Survival and growth of transplanted ramets in different environments, *J. Ecol.*, **73**, 185–198.

Attsat, P.R. and O'Dowd, D.J. (1976) Plant defence guilds. *Science*, **193**, 24–9.

Auclair, A.N. and Cottam, G. (1971) Dynamics of black cherry (*Prunus serotina* Ehrh.) in Southern Wisconsin oakforests, *Ecol. Monogr.*, **41**, 153–77.

Augspurger, C.K. (1983a) Offspring recruitment around tropical trees: changes in cohort distance with time, *Oikos*, **40**, 189–196.

Augspurger, C.K. (1983b) Seed dispersal of the tropical tree *Platypodium elegans*, and the escape of its seedlings from fungal pathogens, *J. Ecol.*, **71**, 759–771.

Augspurger, C.K. (1984) Seedling survival of tropical tree species: interactions of dispersal distance, light-gaps, and pathogens, *Ecology*, **65**, 1705–12.

Austin, M.P. (1985) Continuum concept, ordination methods and niche theory, *Ann. Rev. Ecol. Syst.*, **16**, 39–61.

Bach, C.E. (1980) Effects of plant diversity and time of colonization on a herbivore-plant interaction, *Oecologia*, **44**, 319–26.

Baker, H.G. (1972) Seed weight in relation to environmental conditions in California, *Ecology*, **53**, 997–1010.

Barkham, J.P. (1980) Population dynamics of the wild daffodil (*Narcissus pseudonarcissus*). In Clonal growth, seed reproduction, mortality and the effect of density, *J. Ecol.*, **68**, 607–33.

Barkham, J.P. and Hance, C.E. (1982) Population dynamics of the wild daffodil (*Narcissus pseudonarcissus*). III. Implications of a computer model of 1000 years of population change, *J. Ecol.*, **70**, 323–44.

Barnes, B.V. (1966) The clonal growth habit of American aspens, *Ecology*, **47**, 439–47.

Baskin, J.M. and Baskin, C.C. (1972) Influence of germination date on survival and seed production in a natural population of *Leavenworthia stylosa*, *Am. Midl. Nat.*, **88**, 318–23.

Baskin, J.M. and Baskin, C.C. (1974) Germination and survival in a population of the winter annual *Alyssum alyssoides*, *Can. J. Bot.*, **52**, 2439–45.

Baskin, J.M. and Baskin, C.C. (1979a) Studies on the autecology and population biology of the monocarpic perennial *Grindelia lanceolata*, *Am. Midl. Nat.*, **41**, 290–99.

Baskin, J.M. and Baskin, C.C. (1979b) Studies on the autecology and population biology of the weedy monocarpic perennial, *Pastinaca sativa*, *J. Ecol.*, **67**, 601–10.

Baskin, J.M. and Baskin, C.C. (1980) Ecophysiology of secondary dormancy in seeds of *Ambrosia artemisifolia*, *Ecology*, **61**, 475–80.

Baskin, J.M. and Baskin, C.C. (1983) Seasonal changes in the germination response of buried seeds of *Arabidopsis thaliana* and ecological interpretation, *Bot. Gaz.*, **144**, 540–43.

Baskin, J.M. and Baskin, C.C. (1985) The annual dormancy cycle in buried weed seeds: a continuum, *Bioscience*, **35**, 492–98.

Bazzaz, F.A. and Carlson, R.W. (1979) Photosynthetic contribution of flowers and seeds to reproductive effort of an annual colonizer, *New Phytol.*, **82**, 223–32.

Bazzaz, F.A. and Harper, J.L. (1976) Relationship between plant weight and numbers in mixed populations of *Sinapis alba* (L) Rabenh and *Lepidium sativum* L, *J. Appl. Ecol.*, **13**, 211–16.

Becker, P. and Wong, M. (1985) Seed dispersal, seed predation, and juvenile mortality of *Aglaia* sp. (Meliaceae) in lowland dipterocarp rainforest, *Biotropica*, **17**, 230–37.

Begon, M. and Mortimer, M. (1981) *Population ecology*, Blackwell, Oxford; Sinauer, Sunderland, MA.

Bell, A.D. (1974) Rhizome organization in relation to vegetative spread in *Medeola virginiana*, *J. Arnold Arboretum*, **55**, 458–68.

Bell, A.D. (1979) The hexagonal branching pattern of *Alpinia speciosa* L. (Zingiberaceae), *Ann. Bot.*, **43**, 209–23.

Bell, A.D. and Tomlinson, P.B. (1980) Adaptive architecture in rhizomatous plants, *Bot. J. Linn. Soc.*, **80**, 125–60.

Benjamin, L.R. (1982) Some effects of differing times of seedling emergence, population density and seed size on root-size variation in carrot populations. *J. Agric. Sci. Camb.*, **98**, 537–45.

Bennett, K.D. (1983) Postglacial population expansion of forest trees in Norfolk, UK, *Nature*, **303**, 164–67.

Bentley, S. and Whittaker, J.B. (1979) Effects of grazing by a chrysomelid beetle, *Gastrophysa viridula*, on competition between *Rumex obtusifolius* and *Rumex crispus*, *J. Ecol.*, **67**, 79–90.

Bergh J.P. van den, and Wit, de (1960) Concurrentie tussen thimothee (*Phleum pratense* L.) en reukgras (*Anthoxanthum odoratum* L.), *Jaarboek I.B.S.*, 155–66.

Bergmann, F. (1975) Adaptive acid phosphatase polymorphism in conifer seeds, *Silvae Genet*, **24**, 175–77.

Bergmann, F. (1978) The allelic distribution at an acid phosphatase locus in Norway spruce (*Picea abies*) along similar climatic gradients, *Theor. Appl. Genet.*, **52**, 57–64.

Bierzychudek, P. (1981) Pollinator limitation of plant reproductive effort, *Am. Nat.*, **117**, 838–40.

Bishop, G.F. Davy A.J. and Jeffries, R.L. (1978) Demography of *Hieracium pilosella* in a Breck grassland, *J. Ecol.*, **66**, 615–29.

Black, J.N. (1958) Competition between plants of different initial seed sizes in swards of subterranean clover (*Trifolium subterraneum* L.) with particular reference to leaf area and the light microclimate, *Aust. J. Agric. Res.*, **9**, 299–318.

Black, J.N. (1959) Seed size in herbage legumes, *Herb. Abstr.*, **29**, 235–41.

Blackman, G.E. and Rutter, A.J. (1950) Physiological and ecological studies in the analysis of plant environment: V. An assessment of the factors controlling the distribution of the bluebell (*Scilla nonscripta*) in different communities, *Ann. Bot.*, **14**, 487–520.

Bookman, S.S. (1984) Evidence for selective fruit production in *Asclepias*, *Evolution*, **38**, 72–86.

Borchert, M.I. and Jain, S.K. (1978) The effect of rodent seed predation on four species of California annual grasses, *Oecologia*, **33**, 101–13.

Bosbach, K. Hurka, H. and Haase, R. (1982) The soil seed bank of *Capsella bursa-pastoris* (Cruciferae): its influence on population variability, *Flora*, **172**, 47–56.

Bossema, I. (1979) Jays and oaks: an eco-ethological study of a symbiosis, *Behaviour*, **70**, 1–117.

Bourdeau, P. (1954) Oak seedling ecology determining segregation of species in Piedmont oak–hickory forest, *Ecol. Monogr.*, **24**, 297–320.

Boyd, M. (1986) Ph.D. Thesis, Open University, Milton Keynes, UK.

Bradbury, I.K. (1981) Dynamics, structure and performance of shoot populations of the rhizomatous herb *Solidago canadensis* L. in abandoned pastures, *Oecologia*, **48**, 271–76.

Bradbury, I.K. and Hofstra, G. (1977) Assimilate distribution patterns and carbohydrate concentration changes in organs of *Solidago canadensis* during an annual development cycle, *Can. J. Bot.*, **55**, 1121–7.

Bradshaw, A.D. (1965) Evolutionary significance of phenotypic plasticity in plants, *Adv. Genet.*, **13**, 115–55.

Bradshaw, M.E. and Doody, J.P. (1978a) Plant population studies and their relevance to nature conservation, *Biol. Conserv.*, **14**, 223–42.

Bradshaw, M.E. and Doody, J.P. (1978b) Population-dynamics and biology, Ch. 2. pp. 48–63 in Chapham, A.R. (ed.), *Upper Teesdale, the area and its natural history*. Collins, London.

Brenchley, W.E. and Warrington, K. (1930) The weed seed population of arable soil: I. Numerical estimation of viable seeds and observations on their natural dormancy, *J. Ecol.*, **18**, 235–72.

Brockway, L.H. (1980) *Science and colonial expansion: the role of the British Royal Botanic Gardens*. Academic Press, London and New York.

Brokaw, N.V.L. (1982) The definition of treefall gap and its effect on measures of forest dynamics, *Biotropica*, **14**, 158–60.

Brokaw, N.V.L. (1985) Gap-phase regeneration in a tropical forest, *Ecology*, **66**, 682–7.

Brown, J.H. Davidson, D.W. and Reichman, O.H. (1979) An experimental study of competition between seed-eating desert rodents and ants, *Am. Zoologist*, **19**, 1129–43.

Brown, J.H. Reichman, O.J. and Davidson, D.W. (1979) Granivory in desert ecosystems, *Ann. Rev. Ecol. Syst.*, **10**, 201–27.

Buchanan, G.A. Crowley, R.H. Street, J.E. and McGuire, J.A. (1980) Competition of sicklepod (*Cassia obtusifolia*) and redroot pigweed (*Amaranthus retroflexus*) with cotton (*Gossypium hirsutum*), *Weed Sci.*, **28**, 258–62.

Budd, A.C. Chepil, W.S. and Doughty, J.L. (1954) Germination of weed seeds: II. The influence of crops and fallow on the weed seed population of the soil, *Can. J. Agric. Sci.*, **34**, 18–27.

Burdon, J.J. (1980) Variation in disease resistance within a population of *Trifolium repens*, *J. Ecol.*, **68**, 737–44.

Burdon, J.J. and Chilvers, G.A. (1974) Fungal and insect parasites contributing to niche differentiation in mixed species stands of eucalypt saplings, *Aust. J. Bot.*, **22**, 103–14.

Burdon, J.J. and Chilvers, G.A. (1975) Epidemiology of damping-off disease (*Pythium irregulare*) in relation to density of *Lepidium sativum* seedlings, *Ann. Appl. Biol.*, **81**, 135–43.

Burdon, J.J. and Chilvers, G.A. (1982) Host density as a factor in plant disease ecology, *Ann. Rev. Phytopathol.*, **20**, 143–66.

Burdon, J.J. and Marshall, D.R. (1981) Biological control and the reproductive mode of weeds, *J. Appl. Ecol.*, **18**, 649–58.

Burdon, J.J. and Shattock, R.C. (1980) Disease in plant communities, *Appl. Biol.*, **5**, 145–220.

Burdon, J.J. Groves, R.H. Kaye, P.E. and Speer, S.S. (1984) Competition in mixtures of susceptible and resistant genotypes of *Chondrilla juncea* differentially infected with rust, *Oecologia*, **64**, 199–203.

Callaghan, T.V. (1976) Strategies of growth and population dynamics of plants: 3. Growth and population dynamics of *Carex bigelowii* in an alpine environment, *Oikos*, **27**, 402–13.

Canfield, R.H. (1957) Reproduction and life span of some perennial grasses of southern Arizona, *J. Range Mgmt*, **10**, 199–203.

Canham, C.D. (1985) Suppression and release during canopy recruitment in *Acer saccharum*, *Bull. Torrey Bot. Club*, **112**, 134–45.

Cannell, M.G.R. Rothery, P. and Ford, E.D. (1984) Competition within stands of *Picea sitchensis* and *Pinus contorta*, *Ann. Bot.*, **53**, 349–62.

Caswell, H. (1978a) Predator mediated coexistence: a non-equilibrium model, *Am. Nat.*, **112**, 127–54.

Caswell, H. (1978b) A general formula for the sensitivity of population growth rate to changes in life history parameters, *Theor. Pop. Biol.*, **14**, 215–30.

Caswell, H. (1986) The evolutionary demography of vegetative reproduction. In Jackson, J., Buss, L. & Cook, R.E. (eds) *Population biology and evolution of clonal organisms*. Yale University Press, New Haven, CT.

Caughley, G. and Lawton, J.H. (1981) Plant–herbivore systems. pp. 123–66 in May, R.M. (ed.) *Theoretical ecology*, 2nd Edition. Blackwell, Oxford.

Cavers, P.B. and Steele, M.G. (1984) Patterns of change in seed weight over time on individual plants, *Am. Nat.*, **124**, 324–35.

Champness, S.S. and Morris, K. (1948) The population of buried weed seeds in relation to contrasting pasture and soil types. *J. Ecol.*, **36**, 149–73.

Chancellor, R.J. (1968) The value of biological studies in weed control, *Proc. 9th Br. Weed Control Conf.*, 1129–35.

Chancellor, R.J. (1985) Changes in the weed flora of an arable field cultivated for 20 years, *J. Appl. Ecol.*, **22**, 491–501.

Cheke, A.S. Nanakorn, W. and Yankoses, C. (1979) Dormancy and dispersal of seeds of secondary forest species under the canopy of a primary tropical rain forest in Northern Thailand, *Biotropica*, **11**, 88–95.

Cheplick, G.P. and Quinn, J.A. (1982) *Amphicarpum purshii* and the 'pessimistic strategy' in amphicarpic annuals with subterranean fruit, *Oecologia*, **52**, 327–32.

Cheplick, G.P. and Quinn, J.A. (1983) The shift in aerial/subterranean fruit ratio in *Amphicarpum purshii*: causes and significance, *Oecologia*, **57**, 374–79.

Chesson, P.L. (1986) Environmental variation and the coexistence of species. Ch. 14, pp. 240–56 in Diamond, J. and Case, T.J. (eds) *Community ecology*. Harper and Row, London and New York.

Chew, R.M. and Chew, A.E. (1965) The primary productivity of a desert-shrub (*Larrea tridentata*) community, *Ecol. Monogr.*, **35**, 355–75.

Chilvers, G.A. and Brittain, E.G. (1972) Plant competition mediated by host-specific parasites – a simple model, *Aust. J. Biol. Sci.*, **25**, 748–56.

Chippendale, H.G. and Milton, W.E.J. (1934) On the viable seeds present in the soil beneath pastures, *J. Ecol.*, **22**, 508–31.

Christy, E.J. and Mack, R.N. (1984) Variation in demography of juvenile *Tsuga heterophylla* across the substratum mosaic, *J. Ecol.*, **72**, 75–91.

Clark, D.A. and Clark, D.B. (1984) Spacing dynamics of a tropical rain forest tree: evaluation of the Janzen–Connell model, *Am. Nat.*, **124**, 769–88.

Clay, K. (1984) The effect of the fungus *Atkinsonella hypoxylon* (Clavicipitaceae) on the reproductive system and demography of the grass *Danthonia spicata, New Phytol.*, **98**, 165–75.

Clay, K. Hardy, T.N. and Hammond, A.M. jr. (1985) Fungal endophytes of grasses and their effects on an insect herbivore, *Oecologia*, **66**, 1–5.

Clements, F.E. (1916) *Plant succession: an analysis of the development of vegetation.* Carnegie Institution, Washington DC.

Clements, F.E. Weaver, J.E. and Hanson, H.C. (1929) *Plant competition*, Carnegie, Inst. Wash. Pub., p. 398.

Clements, R.O. and Henderson, I.F. (1979) Insects as a cause of botanical change in swards, *J. Brit. Grassl. Soc. Symp.*, **10**, 157–60.

Clymo, R.S. and Reddaway, E.J.F. (1972) A tentative dry matter balance sheet for the wet blanket bog on Burnt Hill, Moor House NNR, *Moor House Occasional Paper*, No 3, Nature Conservancy, London.

Cody, M.L. (1966) A general theory of clutch size, *Evolution*, **20**, 174–84.

Cody, M.L. (1986) Structural niches in plant communities. Ch. 23, pp. 381–405, in Diamond J. and Case, T.J. (eds) *Community ecology*. Harper and Row, London and New York.

Cole, L.C. (1954) The population consequences of life history phenomena, *Quart. Rev. Biol.*, **29**, 103–37.

Collins, N.J. (1976) Growth and population dynamics of the moss *Polytrichum alpestre* in the maritime Antarctic. Strategies of growth and population dynamics of tundra plants, 2, *Oikos*, **27**, 389–401.

Colosi, J.S. and Cavers, P.B. (1984) Pollination affects percent biomass allocated to reproduction in *Silene vulgaris* (bladder campion), *Am. Nat.*, **124**, 299–306.

Comins, H.N. and Noble, I.R. (1985) Dispersal, variability and transient niches: species coexistence in a uniformly variable environment, *Am. Nat.*, **126**, (in press).

Connell, J.H. (1971) On the role of natural enemies in preventing competitive exclusion in some marine animals and in rain forests. In den Boer, P.J. and Gradwell, G.R. (eds) *Dynamics of populations.* Centre for Agricultural Publishing and Documentation (PUDOC), Wageningen, Netherlands.

Connell, J.H. (1978) Diversity in tropical rainforests and coral reefs, *Science*, **199**, 1302–10.

Connell, J.H. Tracey, J.G. and Webb, L.J. (1984) Compensatory recruitment, growth, and mortality as factors maintaining rain forest tree diversity, *Ecol. Monogr.*, **54**, 141–64.

Connolly, J. (1986) On difficulties with replacement series methodology in mixture experiments, *J. Appl. Ecol.*, **23**, 125–37.

Cook, R.E. (1980) Germination and size-dependent mortality in *Viola blanda*, *Oecologia*, **47**, 115–17.

Cook, R.E. (1983) Clonal plant populations, *Am. Sci.*, **71**, 244–53.

Cook, R.E. (1986) Growth and demography in clonal plants. In Jackson, J., Buss, L., and Cook, R.E. (eds) *Population biology and evolution of clonal organisms*, Yale University Press, New Haven, CT.

Courtney, A.D. (1968) Seed dormancy and field emergence in *Polygonum aviculare*, *J. Appl. Ecol.*, **5**, 675–84.

Crawford-Sidebotham, T.J. (1972) The role of slugs and snails in the maintenance of the cyanogenesis polymorphism of *Lotus corniculatus* L. and *Trifolium repens* L., *Heredity*, **28**, 405–11.

Crawley, M. (1983) *Herbivory*. Blackwell, Oxford and Boston.

Crisp, M.D. and Lange, R.T. (1976) Age structure distribution and survival under grazing of the arid zone shrub *Acacia burkitti*, *Oikos*, **27**, 86–92.

Croat, T.B. (1978) *Flora of Barro Colorado Island*. Stanford University Press, Stanford, CA.

Crow, T.R. (1980) A rainforest chronicle – a 30-year record of change in structure and composition at El-Verde, Puerto-Rico, *Biotropica*, **12**, 42–55.

Culver, D.C. and Beatie, A.J. (1978) Myrmecochory in viola: dynamics of seed-ant interactions in some west Virginian species, *J. Ecol.*, **66**, 53–72.

Curtis, J.T. (1959) *The vegetation of Wisconsin*. University of Wisconsin Press, Madison, Wisconsin.

Darley-Hill, S. and Johnson, W.C. (1981) Acorn dispersal by the blue jay (*Cyanocitta cristata*), *Oecologia*, **50**, 231–32.

Darwin, C. (1859) *The origin of species*. 1st edn, Murray, London.

Davidson, D.W. Samson, D.A. and Inouye, R.S. (1985) Granivory in the Chihuahua desert: interactions within and between trophic levels, *Ecology*, **66**, 486–502.

Davies, M.B. (1981) Quaternary history and the stability of forest communities, Ch. 10, pp. 132–53 in West, D.C. Shugart, H.H., and Botkin, D.B. (eds) *Forest Succession*. Springer-Verlag, New York and Heidelberg.

Davies, M.S. and Snaydon, R.W. (1976) Rapid population differentiation in a mosaic environment. III. Measures of selection pressures, *Heredity*, **36**, 59–66.

Davy, A.J. and Smith, H. (1985) Population differentiation in the life-history characteristics of salt-marsh annuals, *Vegetatio*, **61**, 117–25.

Dayton, P.K. (1971) Competition, disturbance and community organization: the provision and subsequent utilization of space in a rocky intertidal community, *Ecol Monogr.*, **41**, 351–89.

Deevey, E.S. (1947) Life tables for natural populations of animals, *Quart. Rev. Biol.*, **22**, 283–314.

del Moral, R. (1983) Competition as a control mechanism in subalpine meadows, *Am. J. Bot.*, **70**, 232–45.

Dempster, J.P. (1982) The population ecology of the cinnabar moth, *Tyria jacobeae* 1. (Lepidoptera, Arctiidae). *Adv. Ecol. Res.*, **12**, 1–36.

Dempster, J.P. and Lakhani, K.H. (1979) A population model for cinnabar moth and its food plant, ragwort, *J. Anim. Ecol.*, **48**, 143–63.

Diamond, J. and Case, T.J. (eds) (1986) *Community ecology*. Harper and Row, London and New York.

Dirzo, R. and Harper, J.L. (1980) Experimental studies on slug–plant interactions: II. The effect of grazing by slugs on high density monocultures of *Capsella bursa-pastoris* and *Poa annua*, *J. Ecol.*, **68**, 999–1011.

Dodd, A.P. (1940) *The biological campaign against prickly pear.* Commonwealth Prickly Pear Board, pp. 1–177. Government Printer, Brisbane, Australia.

Downs, A.A. (1944) Estimating acorn crops for wild life in the southern Appalachians, *J. Wildl. Mgmt*, **8**, 339–40.

Downs, C. and McQuilkin, W.E. (1944) Seed production of Southern Appalachian oaks, *Journal of Forestry*, **42**, 913–20.

Duckett, J.G. and Duckett, A.R. (1980) Reproductive biology and population dynamics of wild gametophytes of *Equisetum*, *Bot. J. Linn. Soc.*, **80**, 1–40.

Edwards, J. (1984) Spatial pattern and clone structure of the perennial herb, *Aralia nudicaulis* L. (Araliacea), *Bull. Torrey. Bot. Club*, **111**, 28–33.

Eis, S. Garman, E.H. and Ebel, L.F. (1965) Relation between cone production and diameter increment of douglas fir (*Pseudotsuga menziesii* (Mirb.) Franco), grand fir (*Abies grandis* Dougl.), and western white pine (*Pinus monticola* Dougl.), *Can. J. Bot.*, **43**, 1553–9.

Ellenberg, H. (1978) *Vegetation Mitteleuropas mit den Alpen*, 2nd Edition. Ulmer, Stuttgart.

Ennos, R.A. (1981) Detection of selection in populations of white clover (*Trifolium repens* L.), *Biol. J. Linn. Soc.*, **15**, 75–82.

Epling, C. Lewis, H. and Ball, E.M. (1960) The breeding group and seed storage: a study in population dynamics, *Evolution*, **14**, 238–55.

Erith, A.G. (1924) *White clover* (Trifolium repens L.): *a monograph.* Duckworth, London.

Ernst, W.H.O. (1983) Population biology and mineral nutrition of *Anemone nemorosa* with emphasis on its parasitic fungi, *Flora*, **173**, 335–48.

Evans, D.R. Hill, J. Williams, T.A. and Rhodes, I. (1985) Effects of coexistence on the performance of white clover – perennial ryegrass mixtures, *Oecologia*, **66**, 536–39.

Fagerström, T. and Agren, G.I. (1979) Theory for coexistence of species differing in regeneration properties, *Oikos*, **33**, 1–10.

Faille, A. Lemée, G. and Pontailler, J.Y. (1984) Dynamique des clairières d'une forêt inexploitée (réserves biologiques de la forêt de Fontainbleau) I. Origine et état actuel des ouvertures, *Acta Oecologica Oecol. Gener.*, **5**, 35–51.

Falinska, K. (1982) The biology of *Mercurialis perennis* L. polycormones. *Acta Soc. Bot. Poloniae*, **51**, 127–48.

Firbank, L. and Watkinson, A.R. (1985) On the analysis of competition within two-species mixtures of plants, *J. Appl. Ecol.*, **22**, 503–17.

Fisher, R.A. (1930) *The genetical theory of natural selection.* Oxford University Press.

Flor, H.H. (1956) The complimentary genic systems in flax and flax rust, *Adv. Genet.*, **8**, 29–54.

Flor, H.H. (1971) Current status of the gene-for-gene concept, *Ann. Rev. Phytopathol.*, **9**, 275–96.

Flower-Ellis, J.G.K. (1971) Age structure and dynamics in stands of bilberry, *Rapp. Uppsatts. Avdel. Skogsekol*, **9**, 1–108.

Ford, E.D. (1975) Competition and stand structure in some even-aged plant monocultures, *J. Ecol.*, **63**, 311–33.

Fortainier, E.J. (1973) Reviewing the length of the generation period and its shortening, particularly in tulips, *Sci. Hort.*, **1**, 107–16.

Foster, S.A. and Janson, C.H. (1985) The relationship between seed size and establishment conditions in tropical woody plants, *Ecology*, **66**, 773–80.

Fowells, H.A. and Schubert, G.H. (1956) Seeds crops of forest trees in the pine region of California. *USDA Tech. Bull.*, **1150**, 48pp.

Fowler, N.L. (1981) Competition and coexistence in a North Carolina grassland: II. The effects of the experimental removal of species, *J. Ecol.*, **69**, 825–41.

Fox, J.F. (1981) Intermediate levels of soil disturbance maximize alpine plant species diversity, *Nature*, **293**, 564–65.

Fridrikson, S. (1975) *Surtsey, evolution of life on a volcanic island*, Butterworth, London.
Frissell, S.S. (1973) The importance of fire as a natural ecological factor in Itasca State Park, Minnesota, *Quat. Res.*, **3**, 397–407.
Gadgil, M. and Prasad, S.N. (1984) Ecological determinants of life history evolution of two Indian bamboo species, *Biotropica*, **16**, 161–72.
Gadgil, P.M. and Solbrig, O.T. (1972) The concept of *r* and *K* selection. Evidence from some wild flowers and theoretical considerations, *Am. Nat.*, **106**, 14–31.
Garwood, N.C. (1982) Seasonal rhythm of seed germination in a semideciduous tropical forest, pp. 173–85, in Leigh, E.G. Jr. Rand, A.S. and Windsor, D.M. (eds) *The ecology of a tropical forest: seasonal rhythms and long-term changes*. Smithsonian Institution Press, Washington DC.
Garwood, N.C. (1983) Seed germination in a seasonal tropical forest in Panama: A community study, *Ecol. Monogr.*, **53**, 159–81.
Gashwiler, J.S. (1967) Conifer seed survival in a western Oregon clearcut, *Ecology*, **48**, 431–3.
Gatsuk, E. Smirnova, O.V. Vorontzova, L.I. Zaugolnova, L.B. and Zhukova, L.A. (1980) Age states of plants of various growth forms: a review, *J. Ecol.*, **68**, 675–96.
Gauch, H.G. Jr. (1982) *Multivariate analysis in community ecology*. Cambridge University Press, Cambridge.
Gause, G.F. (1934) *The struggle for existence*. Williams and Wilkins, Baltimore.
Gentry, A.H. (1982) Patterns of neotropical plant species diversity, *Evol. Biol.*, **15**, 1–84.
Ginzo, H.D. and Lovell, P.H. (1973) Aspects of the comparative physiology of *Ranunculus bulbosus* and *R. repens*. II. Carbon dioxide assimilation and distribution of photosynthates, *Ann. Bot.*, **37**, 765–76.
Gleason, H.A. (1926) The individualistic concept of the plant association, *Bull. Torrey Bot, Club*, **53**, 7–26.
Gleason, H.A. (1927) Further views on the succession-concept, *Ecology*, **8**, 299–326.
Glier, J.H. and Caruso, J.L. (1973) Low-temperature induction of starch degradation in roots of a biennial weed, *Cryobiology*, **10**, 328–30.
Goldberg, D.E. and Werner, P.A. (1983) Equivalence of competitors in plant communities: A null hypothesis and a field experimental approach, *Amer. J. Bot.*, **70**, 1098–104.
Golubeva, J.N. (1962) Some data on pools of viable seeds in soil under meadow-steppe vegetation (In Russian), *Byull. Mosk. Obshch. Isp. Prir.*, **67**, 76–89.
Gottsberger, G. (1978) Seed dispersal by fish in the inundated regions of Humaita, Amazonia, *Biotropica*, **10**, 170–83.
Gould, S.J. and Lewontin, R.C. (1979) The spandrels of San Marco and the Panglossian paradigm: a critique of the adaptationist programme, *Proc. R. Soc. Lond. B.*, **205**, 581–98.
Grace, J.B. (1985) Juvenile vs. adult competitive abilities in plants: size-dependence in cattails (*Typha*), *Ecology*, **66**, 1630–38.
Grant, M.C. and Antonovics, J. (1978) Biology of ecologically marginal populations of *Anthoxanthum odoratum*: I. Phenetics and dynamics, *Evolution*, **32**, 822–38.
Gray, B. (1972) Economic tropical forest entomology, *Ann. Rev. Entomol.*, **17**, 313–54.
Gray, B. (1975) Size-composition and regeneration of *Araucaria* stands in *New Guinea J. Ecol.*, **63**, 273–89.
Grime, J.P. (1979) *Plant strategies and vegetation processes*. Wiley, Chichester, UK and New York, USA.
Grime, J.P. and Jeffrey, D.W. (1965) Seedling establishment in vertical gradients of sunlight, *J. Ecol.*, **53**, 621–42.
Gross, K.L. (1980) Colonization by *Verbascum thapsus* (Mullein) of an old field in Michigan: experiments on the effects of vegetation, *J. Ecol.*, **68**, 919–28.

Gross, K.L. (1981) Predictions of fate from rosette size in four 'biennial' plant species: *Verbascum thapsus*, *Oenothera biennis*, *Daucus carota*, and *Tragopogon dubius*, *Oecologia*, **48**, 209–13.

Groves, R.H. and Williams, J.D. (1975) Growth of skeleton weed (*Chondrilla juncea* L.) as affected by growth of subterranean clover (*Trifolium subterraneum* L.) and infection by *Puccinia chondrilla* Bubak and Syd, *Aust. J. Agric. Res.*, **26**, 975–83.

Grubb, P.J. (1976) A theoretical background to the conservation of ecologically distinct groups of annuals and biennials in the chalk grassland ecosystem, *Biol. Conserv.*, **10**, 53–76.

Grubb, P.J. (1977) The maintenance of species richness in plant communities. The importance of the regeneration niche, *Biol. Rev.*, **52**, 107–45.

Grubb, P.J. (1986) Problems posed by sparse and patchily distributed species in species-rich plant communities. Ch. 12, pp. 207–25, in Diamond, J. and Case, T.J. (eds) *Community ecology*. Harper and Row, London and New York.

Guittet, J. and Laberche, J.C. (1974) L'implantation naturelle du pin sylvestre sur pelouse xérophile en forêt de Fontainebleau: II. Demographic des graines et des plantules au voisinage des vieux arbres, *Oecol. Plant.*, **9**, 111–30.

Gunnill, F.C. (1980) Demography of the intertidal brown alga *Pelvetia fastigiata* in Southern California, USA *Marine Biol.*, **59**, 169–79.

Hairston, N.G. Tinkle, D.W. and Wilbur, H.M. (1970) Natural selection and the parameters of population growth, *J. Wildl. Mgmt*, **34**, 681–90.

Haizel, K.A. and Harper, J.L. (1973) The effects of density and the timing of removal on interference between barley, white mustard and wild oats, *J. Appl. Ecol..*, **10**, 23–32.

Hall, R.L. (1974) Analysis of the nature of interference between plants of different species: II. Nutrient relations in a Nandi *Setaria* and greenleaf *Desmodium* association with particular reference to potassium, *Aust. J. Agric. Res.*, **25**, 749–56.

Handel, S.N. (1983) Pollination ecology, plant population structure, and gene flow. Ch. 8, pp. 163–211, in Real, L. (ed.) *Pollination biology*. Academic Press, Orlando and London.

Hanf, M. (1974) *Weeds and their seedlings*. BASF, Ipswich, England.

Harberd, D.J. (1961) Observations on population structure and longevity of *Festuca rubra* L., *New Phytol.*, **60**, 184–206.

Harberd, D.J. (1962) Some observations on natural clones in *Festuca ovina*, *New Phytol.*, **61**, 85–100.

Harberd, D.J. (1963) Observations on natural clones of *Trifolium repens* L., *New Phytol.*, **62**, 198–204.

Harcourt, D.G. (1970) Crop life tables as a pest management tool, *Can. Entomol.*, **102**, 950–5.

Hardin, G. (1960) The competitive exclusion principle, *Science*, **131**, 1292–97.

Harlan, J.R. (1976) Diseases as a factor in plant evolution, *Ann. Rev. Phytopathol.*, **14**, 31–51.

Harper, J.L. (1959) The ecological significance of dormancy and its importance in weed control, Proc. 4th Int. Congr. Crop. Prot. (*Hamburg*), 415–20.

Harper, J.L. (1961) Approaches to the study of plant competition, *Soc. Exp. Biol. Symp.*, **15**, 1–39.

Harper, J.L. (1977) *Population biology of plants*. Academic Press, London and New York.

Harper, J.L. (1978) The demography of plants with clonal growth, pp. 27–48 in Freysen, A.H.J. and Woldendorp, J. (eds), *Structure and functioning of plant populations*. North Holland Publ. Co., Amsterdam.

Harper, J.L. (1982) After description, pp. 11–25, in Newman, E.I. (ed.) *The plant community as a working mechanism*. Special Publication of the British Ecological Society No. 1. Blackwell, Oxford and Boston.

Harper, J.L. (1986) Modules, branches and the capture of resources, in Jackson, J.,

Buss, L., and Cook, R.E. (eds) *Population biology and evolution of clonal organisms*. Yale University Press, New Haven, CT.

Harper, J.L. and Gajic, D. (1961) Experimental studies of the mortality and plasticity of a weed, *Weed Res.*, **1**, 91–104.

Harper, J.L. Lovell, P.H. and Moore, K.G. (1970) The shapes and sizes of seeds, *Ann. Rev. Ecol. Syst.*, **1**, 327–56.

Harper, J.L. and Ogden, J. (1970) The reproductive strategy of higher plants: I. The concept of strategy with special reference to *Senecio vulgaris* L., *J. Ecol.*, **58**, 681–98.

Harper, J.L. and McNaughton, I.H. (1962) The comparative biology of closely related species living in the same area: VII. Interference between individuals in pure and mixed populations of *Papaver* spp., *New Phytol.*, **61**, 175–88.

Harper, J.L. and White, J. (1971) The dynamics of plant populations, *Proc. Adv. Study Inst. Dynamics Numbers Popul.* (*Oosterbeek 1970*), 41–63.

Harper, J.L. Williams, J.T. and Sagar, G.R. (1965) The behaviour of seeds in the soil: I. The heterogeneity of soil surfaces and its role in determining the establishment of plants from seed, *J. Ecol.*, **53**, 273–86.

Harris, P. (1984) *Carduus nutans* L., nodding thistle and *C. acanthoides* L., plumeless thistle (Compositae). Ch. 30, pp. 115–26 in Kelleher, J.S. and Hulme, M.A. (eds) Biological control programmes against insects and weeds in Canada, 1969–1980, *Commonw. Inst. Biol. Control Tech. Commun.*

Harris, P. Thompson, L.S. Wilkinson, A.T.S. and Neary, M.E. (1978) Reproductive biology of tansy ragwort, climate and biological control by the cinnabar moth in Canada. *Proc. Fourth Int. Symp. Biol. Control of Weeds*, pp. 163–73.

Hart, R. (1977) Why are biennials so few? *Am. Nat.*, **111**, 792–9.

Hartgerink, A.P. and Bazzaz, F.A. (1984) Seedling-scale environmental heterogeneity influences individual fitness and population structure, *Ecology*, **65**, 198–206.

Hartnett, D.C. and Bazzaz, F.A. (1983) Physiological integration among intraclonal ramets in *Solidago canadensis*, *Ecology*, **64**, 779–88.

Hartnett, D.C. and Bazzaz, F.A. (1985a) The genet and ramet population dynamics of *Solidago canadensis* in an abandoned field, *J. Ecol.*, **73**, 407–13.

Hartnett, D.C. and Bazzaz, F.A. (1985b) The integration of neighbourhood effects by clonal genets in *Solidago canadensis*, *J. Ecol.*, **73**, 415–27.

Hartnett, D.C. and Bazzaz, F.A. (1985c) The regulation of leaf, ramet and genet densities in experimental populations of the rhizomatous perennial *Solidago canadensis*, *J. Ecol.*, **73**, 429–43.

Hartshorn, G.S. (1977) Tree falls and tropical forest dynamics, pp. 617–38 in Tomlinson, P.B. and Zimmerman, M.H. (eds), *Tropical trees as living systems*. Cambridge University Press, Cambridge and New York.

Haslam, S.M. (1978) *River plants*. Cambridge University Press, Cambridge and New York.

Hawthorne, W. and Cavers, P.B. (1976) Population dynamics of the perennial herbs *Plantago major* L. and *P. rugeli Decne*, *J. Ecol.*, **64**, 511–27.

Hayashi, I. and Numata, M. (1971) Ecological studies on the buried seed population in the soil related to plant succession: IV. *Jap. J. Ecol.*, **20**, 243–52.

Heinselman, M.L. (1973) Fire in the virgin forest of the Boundary Waters Canoe Area, Minnesota, *Quat. Res.*, **3**, 329–82.

Hendrix, S.D. (1984) Variation in seed weight and its effects on germination in *Pastinaca sativa* L. (Umbelliferae), *Am. J. Bot.*, **71**, 795–802.

Hett, J.M. (1971) A dynamic analysis of age in sugar maple seedlings, *Ecology*, **52**, 1071–4.

Hett, J.M. and Loucks, O.L. (1971) Sugar maple (*Acer saccharum* Marsh.) seedling mortality, *J. Ecol.*, **59**, 507–20.

Hett, J.M. and Loucks, O.L. (1976) Age structure models of balsam fir and eastern hemlock, *J. Ecol.*, **64**, 1029–44.

Hibbs, D.E. (1979) The age structure of a striped maple population, *Can. J. For. Res.*, **9**, 504–8.

Hibbs, D.E. and Fischer, B.C. (1979) Sexual and vegetative reproduction of striped maple (*Acer pensylvanicum* L.), *Bull. Torrey Bot. Club*, **106**, 222–27.

Hickman, J.C. (1975) Environmental unpredictability and plastic energy allocation strategies in the annual *Polygonum cascadense* (Polygonaceae), *J. Ecol.*, **63**, 689–701.

Hils, M.H. and Vankat, J.L. (1982) Species removals from a first-year old-field plant community, *Ecology*, **63**, 705–11.

Hiroi, T. and Monsi, M. (1966) Dry-matter economy of *Helianthus annuus* communities grown at varying densities and light intensities, *J. Fac. Sci. Tokyo Univ.*, **III-9**, 241–85.

Hirose, T. (1983) A graphical analysis of life history evolution in biennial plants, *Bot. Mag. Tokyo*, **96**, 37–47.

Hirose, T. and Kachi, N. (1982) Critical plant size for flowering in biennials with special reference to their distribution in a sand dune system, *Oecologia*, **55**, 281–4.

Hobbs, R.J. and Mooney, H.A. (1985) Community and population dynamics of serpentine grassland annuals in relation to gopher disturbance, *Oecologia*, **67**, 342–51.

Holm, T. (1899) *Podophyllum peltatum*, a morphological study, *Bot. Gaz.*, **27**, 419–43.

Holmsgaard, E. (1956) Effect of seed-bearing on the increment of European beech (*Fagus sylvatica* L.) and Norway spruce (*Picea abies* (L) Karst), *Proc. Int. Univ. For. Res. Org.*, *12th Congr.*, Oxford, 158–61.

Holt, B.R. (1972) Effect of arrival time on recruitment mortality and reproduction in successional plant populations, *Ecology*, **53**, 668–73.

Horn, H.S. (1966) Measurement of 'overlap' in comparative ecological studies, *Am. Nat.*, **100**, 419–24.

Howe, H.F. and Smallwood, J. (1982) Ecology of seed dispersal. *Ann. Rev. Ecol. Syst.*, **13**, 201–28.

Howe, H.F. (1985) Gomphothere fruits: a critique, *Am. Nat.*, **125**, 853–65.

Hubbell, S.P. (1980) Seed predation and coexistence of tree species in tropical forests, *Oikos*, **35**, 214–29.

Hubbell, S.P. and Foster, R.B. (1985) La estructura espacial en gran escala de un bosque neotropical, *Revista de Biologia Tropical* (in press).

Hubbell, S.P. and Foster, R.B. (1986) Biology, chance, and history and the structure of tropical rain forest tree communities. Ch. 19, pp. 314–29 in Diamond, J. and Case, T.J. (eds) *Community ecology*. Harper and Row, London and New York.

Huntley, B. and Birks, H.J.B. (1983) An atlas of past and present pollen maps of Europe, 0–13,000 years ago. Cambridge University Press, Cambridge and New York.

Huston, M. (1979) A general hypothesis of species diversity, *Am. Nat.*, **113**, 81–101.

Hutchings, M.J. (1978) Standing crop and pattern in pure stands of *Mercurialis perennis* and *Rubus fruticosus* in mixed deciduous woodland, *Oikos*, **31**, 351–7.

Hutchings, M.J. (1979) Weight–density relationships in ramet populations of clonal perennial herbs, with special reference to the −3/2 power law, *J. Ecol.*, **67**, 21–33.

Hutchings, M.J. (1983) Shoot performance and population structure in pure stands of *Mercurialis perennis* L., a rhizomatous perennial herb, *Oecologia*, **58**, 260–4.

Hutchings, M.J. and Bradbury, I.K. (1986) Some ecological perspectives on clonal perennial herbs, *Bioscience*, **36**, 178–82.

Hutchinson, G.E. (1957) The multivariate niche, *Cold. Spr. Harb. Symp. Quant. Biol.*, **22**, 415–21.

Ikusima, I. and Shinozaki, K. (1955) Intraspecific competition among higher plants: II. Growth of duckweed, with a theoretical consideration on the C–D effect, *J. Inst. Polytech. Osaka. Univ.*, Ser. D6, 107–19.

Inouye, R.S. Byers, G.S. and Brown, J.H. (1980) Effects of predation and competition on survivorship, fecundity, and community structure of desert annuals, *Ecology*, **61**, 1344–51.

Inouye, R.S. and Schaffer, W.M. (1981) On the ecological meaning of ratio (de Wit) diagrams in plant ecology, *Ecology*, **62**, 1679–81.

Jalas, J. and Suominen, J. (1973) *Atlas Flora Europaeae Vol. 2.* Committee for mapping the flora of Europe, Helsinki.

Janzen, D.H. (1969) Seed-eaters versus seed size, number, toxicity and dispersal, *Evolution*, **23**, 1–27.

Janzen, D.H. (1970) Herbivores and the number of tree species in tropical forests, *Am. Nat.*, **104**, 501–28.

Janzen, D.H. (1971) Escape of *Casia grandis* L. beans from predators in time and space, *Ecology*, **52**, 964–79.

Janzen, D.H. (1972) Escape in space by *Sterculia apetala* seeds from the bug *Dysdercus fasciatus* in a Costa Rican deciduous forest, *Ecology*, **53**, 350–61.

Janzen, D.H. (1973) Dissolution of mutualism between *Ceropia* and its *Azteca* ants, *Biotropica*, **5**, 15–28.

Janzen, D.H. (1975a) *Ecology of plants in the tropics*. Studies in Biology No. 58, Arnold, London.

Janzen, D.H. (1975b) Behaviour of *Hymenaea coubaril* when its predispersal seed predator is absent, *Science*, **189**, 145–7.

Janzen, D.H. (1976) Why bamboos wait so long to flower, *Ann. Rev. Ecol. Syst.*, **7**, 347–91.

Janzen, D.H. (1980) Specificity of seed-attacking beetles in a Costa Rican deciduous forest, *J. Ecol.*, **68**, 929–52.

Janzen, D.H. (1984) Dispersal of small seeds by big herbivores: foliage is the fruit, *Am. Nat.*, **123**, 338–53.

Janzen, D.H. and Martin, P.S. (1982) Neotropical anachronisms: the fruits the gomphotheres ate, *Science*, **215**, 19–27.

Jeffries, R.L. Davy, A.J. and Rudmik, T. (1981) Population biology of the salt marsh annual *Salicornia europaea* agg., *J. Ecol.*, **69**, 17–31.

Johnson, E.A. (1975) Buried seed populations in the subarctic forest east of Great Slave Lake, Northwest Territories, *Can. J. Bot.*, **53**, 2933–41.

Johnson, M.P. and Cook, S.A. (1968) 'Clutch size' in buttercups, *Am. Nat.*, **102**, 405–11.

Jolliffe, P.A., Minjas, A.N. and Ruenckles, V.C. (1984) A reinterpretation of yield relationships in replacement series experiments, *J. Appl. Ecol.*, **21**, 227–43.

Jong, de, P. Aarssen, L.W. and Turkington, R. (1980) The analysis of contact sampling data, *Oecologia*, **45**, 322–4.

Julien, M.H. Kerr, J.D. and Chan, R.R. (1984) Biological control of weeds: an evaluation, *Protection Ecol.*, **7**, 3–25.

Jurik, T.W. (1985) Differential costs of sexual and vegetative reproduction in wild strawberry populations, *Oecologia*, **66**, 394–403.

Kachi, N. and Hirose, T. (1985) Population dynamics of *Oenothera glazioviana* in a sand-dune system with special reference to the adaptive significance of size-dependent reproduction, *J. Ecol.*, **73**, 887–901.

Kadereit, J.W. and Briggs, D. (1985) Speed of development of radiate and non-radiate plants of *Senecio vulgaris* L. from habitats subject to different degrees of weeding pressure, *New Phytol.*, **99**, 155–69.

Kanzaki, M. (1984) Regeneration in subalpine coniferous forests: I. Mosaic structure and regeneration process in *Tsuga diversifolia* forest, *Bot. Mag. Tokyo*, **97**, 297–311.

Karpov, V.G. (1960) On the species composition of the viable seed supply in the soil of spruce *Vaccinium myrtillus* vegetation (in Russian), *Trudy Mosk. Obshch. Isp. Priorody*, **3**, 131–40.

Kawano, S. (1975) The productive and reproductive biology of flowering plants: II. The concept of life history strategy in plants, *J. Coll. Lib. Arts Toyama Univ. Japan*, **8**, 51–86.

Kawano, S. Hiratsuka, A. and Hayashi, K. (1982) Life history characteristics and survivorship of *Erythronium japonicum*, *Oikos*, **38**, 129–49.

Kawano, S. and Masuda, J. (1980) The productive and reproductive biology of flowering plants: VII. Resource allocation and reproductive capacity in wild populations of *Heloniopsis orientalis* (Thunb.) C. Tanaka (Liliaceae), *Oecologia*, **45**, 307–17.

Kawano, S. and Nagai, Y. (1975) The productive and reproductive biology of flowering plants: I. Life history strategies of three *Allium* species in Japan, *Bot. Mag. Tokyo*, **88**, 281–318.

Kays, S. and Harper, J.L. (1974) The regulation of plant and tiller density in a grass sward, *J. Ecol.*, **62**, 97–105.

Keddy, P.A. (1981) Experimental demography of the sand-dune annual, *Cakile edentula*, growing along an environmental gradient in Nova Scotia, *J. Ecol.*, **69**, 615–30.

Keddy, P.A. (1982) Population ecology on an environmental gradient: *Cakile edentula* on a sand dune, *Oecologia*, **52**, 348–55.

Kellman, M.C. (1970) The viable seed content of some forest soils in coastal British Columbia, *Can. J. Bot.*, **48**, 1383–5.

Kellman, M.C. (1974a) The viable weed seed content of some tropical agricultural soils, *J. Appl. Ecol.*, **11**, 669–77.

Kellman, M.C. (1974b) Preliminary seed budgets for two plant communities in coastal British Columbia, *J. Biogeography*, **1**, 123–33.

Kershaw, K.A. (1962) Quantitative ecological studies from the Landmannahellir, Iceland. II. The rhizome behaviour of *Carex bigelowii* and *Calamagrostis neglecta*, *J. Ecol.*, **50**, 171–9.

Klemow, K.M. and Raynal, D.J. (1981) Population ecology of *Melilotus alba* in a limestone quarry, *J. Ecol.*, **69**, 33–44.

Kohyama, T. and Fujita, N. (1981) Studies on the *Abies* population of Mt. Shimagare. I. Survivorship curve, *Bot. Mag. Tokyo*, 94, 55–68.

Kulman, H.M. (1971) Effects of insect defoliation on growth and mortality of trees, *Ann. Rev. Ent.*, **16**, 289–324.

Lacey, E.P. (1986) The genetic and environmental control of reproductive timing in a short-lived monocarpic species *Daucus carota* (Umbelliferae). *J. Ecol.*, **74**, 73–86.

Lacey, E.P. Real, L. Antonovics, J. and Heckel, D.G. (1983) Variance models in the study of life histories, *Am. Nat.*, **122**, 114–31.

Lack, D. (1954) *Natural regulation of animal numbers*. Clarendon Press, Oxford.

Laine, K.J. and Niemela, P. (1980) The influence of ants on the survival of mountain birches during an *Oporinia autumnata* (Lep., Geometridae) outbreak, *Oecologia*, **47**, 39–42.

Langer, R.H.M. Ryle, S.M. and Jewiss, O.R. (1964) The changing plant and tiller populations of timothy and meadow fescue swards: I. Plant survival and the pattern of tillering, *J. Appl. Ecol.*, **1**, 197–208.

Larson, M.M. and Shubert, G.H (1970) Cone crops of ponderosa pine in central Arizona, including the influence of Abert squirrels. *USDA Forest Serv. Res. Pap.* R.M., **58**, 15pp.

Law, R. (1979) The cost of reproduction in annual meadow grass, *Am. Nat.*, **113**, 3–16.

Law, R. (1983) A model for the dynamics of a plant population containing individuals classified by age and size, *Ecology*, **64**, 224–30.

Law, R. Bradshaw, A.D. and Putwain, P.D. (1977) Life history variation in *Poa annua*, *Evolution*, **31**, 233–46.

Law, R. and Watkinson, A.R. (1986) Response-surface analysis of two-species competition: an experiment on *Phleum arenarium* and *Vulpia fasciculata.*, (in press).

Leak, W.B. (1975) Age distributions in virgin red spruce and northern hardwoods, *Ecology*, **56**, 1451–4.

Leck, M.A. and Graveline, K.J. (1979) The seed bank of a freshwater tidal mash, *Am. J. Bot.*, **66**, 1006–15.

Leverich, W.J. and Levin, D.A. (1979) Age-specific survivorship and reproduction in *Phlox drummondii*, *Am. Nat.*, **113**, 881–903.

Levin, D.A. (1975) Pest pressure and recombination systems in plants, *Am. Nat.*, **109**, 437–51.

Levin, D.A. and Kerster, H.W. (1974) Gene flow in seed plants, *Evol. Biol.*, **7**, 139–220.

Levin, D.A. and Turner, B.L. (1977) Clutch size in the compositae, Ch. 18, pp. 215–22 in Stonehouse, B. and Perrins, C.M. (eds), *Evolutionary ecology*. Macmillan, London.

Liddle, M.J. Budd, C.S.J. and Hutchings, M.J. (1982) Population dynamics and neighbourhood effects in establishing swards of *Festuca rubra*, *Oikos*, **38**, 52–9.

Lieberman, M. John, D.M. and Liberman, D. (1979) Ecology of subtidal algae on seasonally devastated cobble substrates off Ghana, *Ecology*, **60**, 1151–61.

Lieth, H. (1960) Patterns of change within grassland communities, pp. 27–39 in Harper, J.L. (ed.), *The biology of weeds*. Blackwell, Oxford.

Ligon, D.J. (1978) Reproductive interdependence of pinyon jays and pinyon pines, *Ecol. Monogr.*, **48**, 111–26.

Linhart, Y.B. and Tomback, D.F. (1985) Seed dispersal by nutcrackers causes multi-trunk growth form in pines, *Oecologia*, **67**, 107–10.

Lippert, R.D. and Hopkins, H.H. (1950) Study of viable seeds in various habitats in mixed prairies, *Trans. Kansas Acad. Sci.*, **53**, 355–64.

Lloyd, D.G. (1981) Sexual strategies in plants: I. An hypothesis of serial adjustment of maternal investment during one reproductive session, *New Phytol.*, **86**, 69–79.

Lonsdale, W.M. and Watkinson, A.R. (1982) Light and self-thinning, *New Phytol.*, **90**, 431–45.

Lonsdale, W.M. and Watkinson, A.R. (1983) Plant geometry and self-thinning, *J. Ecol.*, **71**, 285–97.

Louda, S.M. (1982a) Limitation of the recruitment of the shrub *Haplopappus squarrosus* (Asteracea) by flower- and seed-feeding insects, *J. Ecol.*, **70**, 43–53.

Louda, S.M. (1982b) Distribution ecology: variation in plant recruitment over a gradient in relation to insect seed predation, *Ecol. Monogr.*, **52**, 25–41.

Louda, S.M. (1983) Seed predation and seedling mortality in the recruitment of a shrub *Haplopappus venetus* (Asteraceae), along a climatic gradient, *Ecology*, **64**, 511–21.

Lovett Doust, J. (1980) Experimental manipulation of patterns of resource allocation in the growth cycle and reproduction of *Smyrnium olusatrum* L. *Biol. J. Linn. Soc.*, **13**, 155–66.

Lovett Doust, J. and Eaton, G.W. (1982) Demographic aspects of flower and fruit production in bean plants, *Phaseolus vulgaris* L., *Am. J. Bot.*, **69**, 1156–64.

Lovett Doust, L. (1981a) Population dynamics and local specialization in a clonal perennial (*Ranunculus repens*). I. The dynamics of ramets in contrasting habitats, *J. Ecol.*, **69**, 743–55.

Lovett Doust, L. (1981b) Population dynamics and local specialization in a clonal perennial (*Ranunculus repens*). II. The dynamics of leaves, and a reciprocal transplant–replant experiment, *J. Ecol.*, **69**, 757–68.

MacArthur, R.H. (1972) *Geographical ecology*. Harper Row, New York.

MacArthur, R.H. and Levins, R. (1967) The limiting similarity, convergence and divergence of coexisting species, *Am. Nat.*, **101**, 377–85.

MacArthur, R.H. and Wilson, E.O. (1967) *The theory of island biogeography*. Princeton University Press, Princeton, N.J.

Mack, R.N. (1976) Survivorship of *Cerastium atrovirens* at Aberffraw, Anglesey, *J. Ecol.*, **64**, 309–12.

Mack, R.N. and Harper, J.L. (1977) Interference in dune annuals: spatial pattern and neighbourhood effects, *J. Ecol.*, **65**, 345–64.

Mack, R.N. and Pyke, D.A. (1984) The demography of *Bromus tectorum*: The role of microclimate, grazing and disease, *J. Ecol.*, **72**, 731–48.

Maddox, G.D. and Root, R.B. (1987) Resistance to 16 diverse species of herbivorous

insects within a population of goldenrod, *Solidago altissima*: genetic variation and heritability. *Oecologia* (in press).
Major, J. and Pyott, W.T. (1966) Buried viable seeds in California bunchgrass sites and their bearing on the definition of a flora, *Vegetatio Acta Geobotanica*, **13**, 253–82.
Marchand, P.J. and Roach, D.A. (1980) Reproductive strategies of pioneering alpine species: seed production, dispersal, and germination, *Arc. Alp. Res.*, **12**, 137–46.
Marks, M. and Prince, S. (1981) Influence of germination date on survival and fecundity in wild lettuce *Lactuca serriola*, *Oikos*, **36**, 326–30.
Marks, P.L. (1974) The role of pin cherry (*Prunus pennsylvanica* L.) in the maintenance of stability in northern hardwood ecosystems, *Ecol. Monogr.*, **44**, 73–88.
May, R.M. (1975) Some notes on estimating the competition matrix α, *Ecology*, **56**, 737–41.
Mayer, A.M. and Polijakof-Mayber, A. (1975) *The germination of seeds*. Pergamon, Oxford.
Maynard Smith, J. Burian, R. Kauffman, S. Alberch, P. Campbell, J. Goodwin, B.C. Lande, R. Raup, D. and Wolpert, L. (1985) Developmental constraints and evolution, *Quart. Rev. Biol.*, **60**, 265–87.
McBrien, H. Harmsen, R. and Crowder, A. (1983) A case of insect grazing affecting plant succession, *Ecology*, **64**, 1035–9.
Mead, R. and Willey, R.W. (1980) The concept of a 'Land Equivalent Ratio' and advantages in yields from intercropping, *Exp. Agr.*, **16**, 217–28.
Meijden, E. van der. (1979) Herbivore exploitation of a fugitive plant species: local survival and extinction of the cinnabar moth and ragwort in a heterogeneous environment, *Oecologia*, **42**, 307–23.
Meijden, E. van der. Bemmelen, M. van. Kooi, R. and Post, B.J. (1984) Nutritional quality and chemical defense in the ragwort–cinnabar moth interaction, *J. Anim. Ecol.*, **53**, 443–53.
Meijden, E. van de. Jong, T.J. de. Klinkhamer, P.G.L. and Kooi, R.E. (1985) Temporal and spatial dynamics of biennial plants. In Haeck, J. and Woldendorp, J.W. (eds) *Structure and functioning of plant populations II: Phenotypic and genotypic variation in plant populations*. North-Holland, Amsterdam and Oxford.
Mellanby, K. (1968) The effects of some mammals and birds on regeneration of oak, *J. Appl. Ecol.*, **5**, 359–66.
Michon, G. (1983) Village forest gardens in West Java, Ch. 2, pp. 13–24 in Huxley, P.A. (ed.) *Plant research and agroforestry*. International Council for Research in Agroforestry, Nairobi, Kenya.
Milton, W.E.J. (1939) The occurrence of buried viable seeds in soils at different elevations and on a salt marsh, *J. Ecol.*, **27**, 149–59.
Mohler, C.L. Marks, P.L. and Sprugel, D.G. (1978) Stand structure and allometry of trees during self-thinning of pure stands, *J. Ecol.*, **66**, 599–614.
Morris, E.C. and Myerscough, P.J. (1984) The interaction of density and resource levels in monospecific stands of plants: a review of hypotheses and evidence, *Aust. J. Ecol.*, **9**, 51–62.
Mortimer, A.M. (1983) On weed demography, Ch. 2, pp. 3–40, in Fletcher, W.W. (ed.) *Recent advances in weed control*. Commonwealth Agricultural Bureau, Farnham Royal.
Mortimer, A.M. (1984) Population ecology and weed science, Ch. 18, pp. 363–88, in Dirzo, R. and Sarukhan, J. *Perspectives on plant population ecology*. Sinauer, Sunderland, MA.
Muir, P.S. and Lotan, J.E. (1985) Disturbance history and serotiny of *Pinus contorta* in Western Montana, *Ecology*, **66**, 1658–68.
Mukerjii, S.K. (1936) Contribution to the autecology of *Mercurialis perennis* L., *J. Ecol.*, **24**, 38–81.
Myers, J.H. (1980) Is the insect or the plant the driving force in the cinnabar moth–tansy ragwort system?, *Oecologia*, **47**, 16–21.
Murphy, G.I. (1968) Pattern in life history and the environment, *Am. Nat.*, **102**, 390–404.

Myers, J.H. Monro, J. and Murray, N. (1981) Egg clumping, host plant selection and population regulation in *Cactoblastis cactorum (Lepidoptera), Oecologia*, **51**, 7–13,

Naka, K. and Yoneda, T. (1984) Community dynamics of evergreen broadleaf forests in Southwestern Japan. III. Revegetation in gaps in an evergreen oak forest, *Bot. Mag. Tokyo*, **97**, 275–86.

Naylor, R.E.L (1972) Aspects of the population dynamics of the weed *Alopecurus myosuroides* Huds. in winter cereal crops, *J. Appl. Ecol.*, **9**, 127–39.

Nelson, J.F. and Chew, R.M. (1977) Factors affecting seed reserves in the soil of a Mojave desert ecosystem, Rock valley, Nye County, Nevada, *Am. Midl. Nat.*, **97**, 300–20.

Newell, S.J. (1982) Translocation of ^{14}C photoassimilate in two stoloniferous *Viola* species, *Bull. Torrey. Bot. Club*, **109**, 306–17.

Newell, S.J. Solbrig, O.T. and Kincaid, D.T. (1981) Studies on the population biology of the genus *Viola*. III. The demography of *Viola blanda* and *V. pallens*, *J. Ecol.*, **69**, 997–1016.

Niemela, P. Tuomi, J. and Haukioja, E. (1980) Age-specific resistance in trees: defoliation of tamaracks (*Larix laricina*) by larch bud moth (*Zeiraphera improbana*) (Lep., Tortricidae), *Rep. Kevo Subarctic Res. Stat.*, **16**, 49–57.

Ng, F.S.P. (1977) Strategies of establishment in Malayan forest trees, Ch. 5, pp. 129–62 in Tomlinson, P.B. and Zimmerman, M.H. (eds), *Tropical trees as living systems*. Cambridge University Press, Cambridge and New York.

Noble, J.C. Bell, A.D. and Harper, J.L. (1979) The population biology of plants with clonal growth: I. The morphology and structural demography of *Carex arenaria*, *J. Ecol.*, **67**, 983–1008.

Noble, J.C. and Marshall, C. (1983) The population biology of plants with clonal growth. II. The nutrient strategy and modular physiology of *Carex arenaria*, *J. Ecol.*, **71**, 865–77.

O'Dowd, D.J. and Hay, M.E. (1980) Mutualism between harvester ants and a desert ephemeral: seed escape from rodents, *Ecology*, **61**, 531–40.

Odum, E.P. (1971) *Fundamentals of ecology*. 3rd edn. W.B. Saunders and Co., Philadelphia.

Odum, S. (1978) *Dormant seeds in Danish ruderal soils*. Horsholm Arboretum, Denmark.

Ogden, J. (1974) The reproductive strategy of higher plants: II. The reproductive strategy of *Tussilago farfara* L., *J. Ecol.*, **62**, 291–324.

Oinonen, E. (1967a) Sporal regeneration of ground pine (*Lycopodium complanatum* L.) in southern Finland in the light of the size and age of its clones, *Acta For. Fenn.*, **83**, 76–85.

Oinonen, E. (1967b) The correlation between the size of Finnish bracken (*Pteridium aquilinum* (L.) Kuhn) clones and certain periods of site history, *Acta For. Fenn.*, **83**, 1–51.

Oinonen, E. (1969) The time-table of vegetative spreading of the lily-of-the-valley (*Convallaria majalis* L.) and the wood small-reed (*Calamagrostis epigeios* (L.) Roth) in southern Finland, *Acta For. Fenn.*, **97**, 1–35.

Oomes, M.J. and Elberse, W.Th. (1976) Germination of six grassland herbs in microsites with different water contents, *J. Ecol.*, **64**, 743–55.

Oosting, H.J. and Humphries, M.E. (1940) Buried viable seed in a successional series of old field and forest soils, *Bull. Torrey Bot. Club*, **67**, 253–73.

Open University (1981) *Evolutionary Ecology*. Unit 11 of S364 Evolution: Science, a third level course. Open University Press, Milton Keynes.

Pacala, S.W. and Silander, J.A. Jr. (1985) Neighbourhood models of plant population dynamics. I. Single-species models of annuals, *Am. Nat.*, **125**, 385–411.

Paine, R.T. (1979) Disaster catastrophe and local persistence of the sea palm *Postelsia palmaeformis*, *Science*, **205**, 685–7.

Palmblad, I.G. (1968) Competition studies on experimental populations of weed with emphasis on the regulation of population size, *Ecology*, **49**, 26–34.

Palmer, J.H. (1958) Studies on the behaviour of the rhizome of *Agropyron repens* (L.) Beauv. I. The seasonal development and growth of the parent plant and rhizome, *New Phytol.*, **57**, 145–59.

Parker, M.A. (1985) Size-dependent herbivore attack and the demography of an arid grassland shrub, *Ecology*, **66**, 850–60.

Peñalosa, J. (1983) Shoot dynamics and adaptive morphology of *Ipomoea phillomega* (Vell.) House (Convolvulaceae), a tropical rainforest liana, *Ann. Bot.*, **52**, 737–54.

Perkins, D.F. (1968) Ecology of *Nardus stricta* L: I. Annual growth in relation to tiller phenology, *J. Ecol.*, **56**, 633–46.

Pickett, S.T.A. and White, P.S. (1985) *The ecology of natural disturbance and patch dynamics*. Academic Press, Orlando, Fl.

Pielou, E.C. (1977) *Mathematical ecology*. Wiley, New York and Chichester.

Pinder, III. J.E. (1975) Effects of species removal on an old-field plant community, Ecology, **56**, 747–51.

Pinero, D. Martinez-Ramos, M. and Sarukhán, J. (1984) A population model of *Astrocaryum mexicanum* and a sensitivity analysis of its finite rate of increase, *J. Ecol.*, **72**, 977–91.

Pinero, D. and Sarukhán, J. (1982) Reproductive behaviour and its individual variability in a tropical palm, *Astrocaryum mexicanum*, *J. Ecol.*, **70**, 461–72.

Piper, J.G., Charlesworth, B. and Charlesworth, D. (1984) A high rate of self-fertilization and increased seed fertility of homostyle primroses. *Nature*, **310**, 50–51.

Pitelka, L.F. Hansen, S.B. and Ashmun, J.W. (1985) Population biology of *Clintonia borealis*. I. Ramet and patch dynamics, *J. Ecol.*, **73**, 169–83.

Pitelka, L.F. and Ashmun, J.W. (1986) The physiology and ecology of connections between ramets in clonal plants. In Jackson, J., Buss, L., and Cook, R.E. (eds) *Population biology and evolution of clonal organisms*. Yale University Press, New Haven, CT.

Pitelka, L.F. Stanton, D.S. and Peckenham, M.O. (1980) Effects of light and density on resource allocation in a forest herb, *Aster acuminatus* (Compositae), *Am. J. Bot.*, **67**, 942–8.

Platt, W.J. (1975) The colonization and formation of equilibrium plant species associations on badger disturbances in a tall-grass prairie, *Ecol. Monogr.*, **45**, 285–305.

Platt, W.J. and Weiss, I.M. (1977) Resource partitioning and competition within a guild of fugitive prairie plants, *Am. Nat.*, **111**, 479–513.

Platt, W.J. and Weiss, I.M. (1985) An experimental study of competition among fugitive prairie plants, *Ecology*, **66**, 708–20.

Porter, J.R. (1983a) A modular approach to analysis of plant growth. 1. Theory and principles, *New Phytol.*, **94**, 183–90.

Porter, J.R. (1983b) A modular approach to analysis of plant growth. 2. Methods and results, *New Phytol.*, **94**, 191–200.

Primack, R.B. (1979) Reproductive effort in annual and perennial species of *Plantago* (Plantaginaceae), *Am. Nat.*, **114**, 51–62.

Puckridge, D.W. and Donald, C.M. (1967) Competition among wheat plants sown at a wide range of densities, *Aust. J. Agric. Res.*, **17**, 193–211.

Purseglove, J.W. (1968) *Tropical crops. Dicotyledons*. Longman, London.

Putz, F.E. (1983) Treefall pits and mounds, buried seeds, and the importance of soil disturbance to pioneer trees on Barro Colorado Island, Panama, *Ecology*, **64**, 1069–74.

Queller, D.C. (1983) Sexual selection in a hermaphroditic plant, *Nature*, **305**, 706–7.

Rabotnov, T.A. (1964) The biology of monocarp perennial meadow plants, *Bull. Moscow Soc. Nature*, **69**, 57–66. Russian Translation Service 8739, British Library.

Rabotnov, T.A. (1978a) Structure and method of studying coenotic populations of perennial herbaceous plants, *Sov. J. Ecol.*, **9**, 99–105.

Rabotnov, T.A. (1978b) On coenopopulations of plants reproducing by seeds, pp. 1–26 in Freysen, A.H.J. and Woldendrop, J. (eds), *Structure and functioning of plant populations*. North Holland Publ. Co., Amsterdam.

Rackham, O. (1976) *Trees and woodland in the British landscape*. Dent, London.

Ramakrishnan, P.S. (1984) The science behind rotational bush fallow agriculture system (jhum), *Proc. Indian Acad. Sci. (Plant Sci.)*, **93**, 379–400.

Reinartz, J.A. (1984) Life history variation of common mullein (*Verbascum thapsus*). I. Latitudinal differences in population dynamics and timing of reproduction, *J. Ecol.*, **72**, 897–912.

Riley, J. (1984) A general form of the 'Land Equivalent Ratio', *Exp. Agr.*, **20**, 19–29.

Roberts, H.A. (1970) Viable weed seeds in cultivated soils, *Rep. Natn. Veg. Res. Stn*, (1969) 23–38.

Roberts, H.A. (1979) Periodicity of seedling emergence and seed survival in some Umbelliferae, *J. Appl. Ecol.*, **16**, 195–201.

Roberts, H.A. (1981) Seed banks in the soil, *Adv. Appl. Biol.*, **6**, 1–55.

Roberts, H.A. and Feast, P.M. (1973) Emergence and longevity of seeds of annual weeds in cultivated and undisturbed soil, *J. Appl. Ecol.*, **10**, 133–43.

Roberts, H.A. and Ricketts, M.E. (1979) Quantitative relationships between the weed flora after cultivation and the seed population in the soil, *Weeds Res.*, **19**, 269–75.

Robinson, R.G. (1949) Annual weeds, their viable seed population in the soil, and their effect on yields of oats, wheat and flax, *Agron. J.*, **41**, 513–18.

Rockwood, L.L. (1985) Seed weight as a function of life form, elevation and life zone in neotropical forests, *Biotropica*, **17**, 32–9.

Rogan, P.G. and Smith, D.E. (1974) Patterns of translocation of 14-C-labelled assimilates during vegetative growth of *Agropyron repens* (L.) Beauv., *Z. Pflanzenphysiol.*, **73**, 405–14.

Root, R.B. (1973) Organisation of a plant–arthropod association in simple and diverse habitats: the fauna of collards, *Ecol. Monogr.*, **43**, 95–124.

Roughton, R.D. (1962) A review of literature on dendrochronology and age determination of woody plants, *Colo. Dep. Game Fish Tech. Bull.*, **15**, 99pp.

Roughton, R.D. (1972) Shrub age structures on a mule deer winter range in Colorado, *Ecology*, **53**, 615–25.

Rousseau, S. and Loiseau, P. (1982) Structure et cycle de dévelopment des peuplements à *Cytisus scoparius* L. dans la chaine des Puys, *Acta Oecol. Oecol. Applic.*, **3**, 155–68.

Runkle, J.R. (1982) Patterns of disturbance in some old-growth mesic forests of eastern North America, *Ecology*, **63**, 1533–46.

Ryle, G.J.A. Powell, C.E. and Gordon, A.J. (1981) Patterns of 14-C-labelled assimilate partitioning in red and white clover during vegetative growth, *Ann. Bot.*, **47**, 505–14.

Salisbury, E.J. (1942) *The reproductive capacity of plants*. Bell and Sons, London.

Salzman, A.G. (1985) Habitat selection in a clonal plant, *Science*, **228**, 603–4.

Salzman, A.G. and Parker, M.A. (1985) Neighbors ameliorate local salinity stress for a rhizomatous plant in a heterogeneous environment, *Oecologia*, **65**, 273–7.

Sarukhán, J. (1974) Studies on plant demography: *Ranunculus repens* L., *R bulbosus* L. and *R. acris* L: II. Reproductive strategies and seed population dynamics, *J. Ecol.*, **62**, 151–77.

Sarukhán, J. (1977) Studies on the demography of tropical trees, Ch. 6, pp. 163–82 in Tomlinson, P.B. and Zimmerman, M.H. (eds), *Tropical trees as living systems*, Cambridge University Press, Cambridge and New York.

Sarukhán, J. (1980) Demographic problems in tropical systems, Ch. 8, pp. 161–88 in Solbrig, O.T. (ed.), *Demography and evolution in plant populations*. Blackwell, Oxford; University of California Press, California.

Sarukhán, J. and Gadgil, M. (1974) Studies on plant demography: *Ranunculus repens* L., *R. bulbosus* L. and *R. acris* L: III. A mathematical model incorporating multiple modes of reproduction, *J. Ecol.*, **62**, 921–36.

Sarukhán, J. and Harper, J.L. (1973) Studies on plant demography: *Ranunculus repens* L., *R. bulbosus* L. and *R. acris* L: I. Population flux and survivorship, *J. Ecol.*, **61**, 675–716.

Saxena, K.G. and Ramakrishnan, P.S. (1984) Herbaceous vegetation development and weed potential in slash and burn agriculture (jhum) in N.E. India, *Weed Res.*, **24**, 135–42.

Schaffer, W. (1974) Optimal reproductive effort in fluctuating environments, *Am. Nat.*, **108**, 783–90.

Schaffer, W.M. and Gadgil, M.D. (1975) Selection for optimal life histories in plants, Ch. 6, pp. 142–56 in Cody, M.L. and Diamond, J.M. (eds), *Ecology and evolution of communities*. Belknap Press, Cambridge Mass. and London, England.

Schaffer, W.M. and Schaffer, M.V. (1977) The adaptive significance of variations in reproductive habit in the *Agavaceae*, Ch. 22, pp. 26–276 in Stonehouse, B. and Perrins, C.M. (eds), *Evolutionary ecology*. Macmillan, London.

Schaffer, W.M. and Schaffer, M.V. (1979) The adaptive significance of variations in reproductive habit in the Agavaceae: II.—Pollinator foraging behaviour and selection for increased reproductive expenditure, *Ecology*, **60**, 1051–69.

Schellner, R.H. Newell, S.J. and Solbrig, O.T. (1982) Studies of the population biology of the genus *Viola*. IV. Spatial patterns of ramets and seedlings in stoloniferous species, *J. Ecol.*, **70**, 273–90.

Schemske, D.W. (1978) Evolution of reproductive characteristics in *Impatiens* (Balsaminaceae): the significance of cleistogamy and chasmogamy, *Ecology*, **59**, 596–613.

Schemske, D.W. (1984) Population structure and local selection in *Impatiens pallida* (Balsaminaceae), a selfing annual, *Evolution*, **38**, 817–32.

Schneider, K. (1985) Tumors caused by wasps imperiling pin oak trees in Queens and on Long Island, *New York Times*, 24 Oct., 1985.

Scott, P.R. Johnson, R. Wolfe, M.S. Lowe, H.J.B. and Bennett, F.G.A. (1980) Host-specificity in cereal parasites in relation to their control, *Appl. Biol.*, **5**, 349–93.

Selman, M. (1970) The population dynamics of *Avena fatua* (wild oats) in continuous spring barley: desirable frequency of spraying with tri-allate. *Proc. 10th Brit. Weed Control Conf.*, pp. 1176–88.

Sharitz, R.R. and McCormick, J.F. (1975) Population dynamics of two competing annual plant species, *Ecology*, **54**, 723–40.

Shaw, M.W. (1968) Factors affecting the natural regeneration of sessile oak (*Quercus petraea*) in North Wales. I. A preliminary study of acorn production, viability and losses, *J. Ecol.*, **56**, 565–83.

Sheldon, J.C. and Burrows, F.M. (1973) The dispersal effectiveness of the achenepappus units of selected compositae in steady winds with convection, *New Phytol.*, **72**, 665–75.

Shmida, A. and Ellner, S.P. (1984) Coexistence of plants with similar niches, *Vegetatio*, **58**, 29–55.

Shugart, H.H.J. and West, D.C. (1977) Development of an appalachian deciduous forest succession model and its application to assessment of the impact of chestnut blight, *J. Env. Mgmt.*, **5**, 161–80.

Silander, J.A. Jr. (1983) Demographic variation in the Australian desert cassia under grazing pressure, *Oecologia*, **60**, 227–33.

Silander, J.A. and Antonovics, J. (1982) Analysis of interspecific interactions in a coastal plant community – a perturbation approach, *Nature*, **298**, 577–60.

Silander, J.A. Jr. and Pacala, S.W. (1985) Neighbourhood predictors of plant performance, *Oecologia*, **66**, 256–63.

Silvertown, J.W. (1980a) The evolutionary ecology of mast seeding in trees, *Biol. J. Linn. Soc.*, **14**, 235–50.

Silvertown, J.W. (1980b) Leaf-canopy induced seed dormancy in a grassland flora, *New Phytol.*, **85**, 109–18.

Silvertown, J. (1980c) The dynamics of a grassland ecosystem: botanical equilibrium in the Park Grass Experiment, *J. Appl. Ecol.*, **17**, 491–504.

Silvertown, J.W. (1981a) Microspatial heterogeneity and seedling demography in species rich grassland, *New Phytol.*, **88**, 117–28.

Silvertown, J.W. (1981b) Seed size lifespan and germination date as co-adapted features of plant life history, *Am. Nat.*, **118**, p. 860–64.

Silvertown, J.W. (1982) *Introduction to plant population ecology*, 1st Edition. Longman, Harlow and New York.

Silvertown, J. (1983) Why are biennials sometimes not so few?, *Am. Nat.*, **121**, 448–53.

Silvertown, J.W. (1984) Phenotypic variety in seed germination behavior: the ontogeny and evolution of somatic polymorphism in seeds, *Am. Nat.*, **124**, 1–16.

Silvertown, J. (1985) History of a latitudinal diversity gradient: woody plants in Europe 13,000–1000 years B.P. *J. Biogeog.*, **12**, 519–25.

Silvertown, J. (1986) 'Biennials': reply to Kelly, *Am. Nat.*, **127**, 721–4.

Sobey, D.G. and Barkhouse, P. (1977) The structure and rate of growth of the rhizomes of some forest herbs and dwarf shrubs of the New Brunswick–Nova Scotia border region, *Can. Field Nat.*, **91**, 377–83.

Smith, A.P. and Palmer, J.O. (1976) Vegetative reproduction and close packing in a successional plant species, *Nature*, **261**, 232–3.

Smith, F.G. and Thornberry, G.O. (1951) The tetrazolium test and seed viability, *Proc. Ass. Off. Seed Analysts N. Am.*, **41**, 105–8.

Snell, T.W. and Burch, D.G. (1975) The effects of density on resource partitioning in *Chamaesyce hirta* (Euphorbiaceae), *Ecology*, **56**, 742–6.

Snoad, B. (1981) Plant form, growth rate and relative growth rate compared in conventional, semi-leafless and leafless peas, *Scient. Hort.*, **14**, 9–18.

Soane, I.D. and Watkinson, A.R. (1979) Clonal variation in populations of *Ranunculus repens*, *New Phytol.*, **82**, 537–73.

Sohn, J.J. and Policansky, D. (1977) The costs of reproduction in the mayapple *Podophyllum peltatum* (Berberidaceae), *Ecology*, **58**, 1366–74.

Solbrig, O.T. and Simpson, B.B. (1974) Components of regulation of a population of dandelions in Michigan, *J. Ecol.*, **62**, 473–86.

Solbrig, O.T. and Simpson, B.B. (1977) A garden experiment on competition between biotypes of the common dandelion (*Taraxacum officinale*), *J. Ecol.*, **65**, 427–30.

Stakman, E.C. Kempton, F.E. and Hutton, D. (1927) The common barberry and black stem rust, *USDA Farmers Bulletin* No. 1544, 28 pp.

Stanton, M.L. (1984) Developmental and genetic sources of seed weight variation in *Raphanus raphanistrum* L. (Brassicaceae), *Am. J. Bot.*, **71**, 1090–8.

Steenis, C.G.G.J. van. (1981) *Rheophytes of the world*, Sijthoff & Noordhoff, Alphen aan den Rijn, Netherlands.

Stearns, F.W. (1949) Ninety years change in a northern hardwood forest in Wisconsin, *Ecology*, **30**, 350–8.

Stearns, S.C. (1976) Life history tactics: a review of the ideas, *Quart. Rev. Biol.*, **51**, 3–47.

Stearns, S.C. (1977) The evolution of life history traits, *Ann. Rev. Ecol. Syst.*, **8**, 145–71.

Stephenson, A.G. (1979) An evolutionary examination of the floral display of *Catalpa speciosa* (Bignoniaceae), *Evolution*, **33**, 1200–9.

Stephenson, A.G. (1981) Flower and fruit abortion: proximate causes and ultimate functions, *Ann. Rev. Ecol. Syst.*, **12**, 253–79.

Stowe, L.G. and Wade, M.J. (1979) The detection of small-scale patterns in vegetation, *J. Ecol.*, **67**, 1047–64.

Summerhayes, V.S. (1968) *Wild orchids of Britain*. 2nd edn, Collins, London.

Sutherland, S. and Delph, L.F. (1984) On the importance of male fitness in plants: patterns of fruit set, *Ecology*, **65**, 1093–104.

Swain, A.M. (1973) A history of fire and vegetation in northeastern Minnesota as recorded in lake sediments, *Quat. Res.*, **3**, 383–96.

Symonides, E. (1977) Mortality of seedlings in natural psammophyte populations, *Ekol. Pol.*, **25**, 635–51.

Symonides, E. (1978) Numbers, distribution and specific composition of diaspores in the soils of the plant association spergulo-corynephoretum, *Ekol. Pol.*, **26**, 111–22.

Symonides, E. (1979a) The structure and population dynamics of psammophytes of inland dunes: I. Populations of initial stages, *Ekol. Pol.*, **27**, 3–37.

Symonides, E. (1979b) The structure and population dynamics of psammophytes on inland dunes: II. Loose-sod populations, *Ekol. Pol.*, **27**, 191–234.

Symonides, E. (1979c) The structure and population dynamics of psammophytes on inland dunes: III. Populations of compact psammophyte communities, *Ekol. Pol.*, **27**, 235–57.

Symonides, E. (1983a) Population size regulation as a result of intra-population interactions. II. Effect of density on the survival of individuals of *Erophila verna* (L.) C.A.M. *Ekol. Polska*, **31**, 839–81.

Symonides, E. (1983b) Population size regulation as a result of intra-population interactions. III. Effect of density on the growth rate, morphological diversity and fecundity of *Erophila verna* (L.) C.A.M. individuals, *Ekol. Polska*, **31**, 883–912.

Symonides, E. (1984) Population size regulation as a result of intra-population interactions. I. Effect of *Erophila verna* (L.) C.A.M. population density on the abundance of the new generation of seedlings. Summing-up and conclusions, *Ekol. Polska*, **32**, 557–80.

Tamm, C.O. (1956) Further observations on the survival and flowering of some perennial herbs: 7. *Oikos*, **7**, 274–92.

Tamm, C.O. (1972a) Survival and flowering of some perennial herbs: II. The behaviour of some orchids on permanent plots, *Oikos*, **23**, 23–8.

Tamm, C.O. (1972b) Survival and flowering of perennial herbs: III. The behaviour of *Primula veris* on permanent plots, *Oikos,* **23**, 159–66.

Tansley, A.G. (1917) On competition between *Galium saxatile* L. (*G. hercynium* Weig.) and *Galium sylvestre* Poll. (*G. asperum* Schreb.) on different types of soil, *J. Ecol.*, **5**, 173–9.

Temple, S.A. (1977) Plant–animal mutualism: coevolution with the dodo leads to near extinction of plant, *Science*, **197**, 885–6.

Templeton, A.R. and Levin, D.A. (1979) Evolutionary consequences of seed pools, *Am. Nat.*, **114**, 232–49.

Thomas, A.G. and Dale, H.M. (1974) The role of seed reproduction in the dynamics of established populations of *Hieracium floribundum* and a comparison with that of vegetative reproduction, *Can. J. Bot.*, **53**, 3022–31.

Thompson, D.A. and Beattie, A.J. (1981) Density-mediated seed and stolon production in *Viola* (Violaceae), *Am. J. Bot.*, **68**, 383–8.

Thompson, K. (1986) Small-scale heterogeneity in the seed bank of an acidic grassland, *J. Ecol.*, **74**, 733–8

Thompson, K. and Grime, J.P. (1979) Seasonal variation in seed banks of herbaceous species in ten contrasting habitats. *J. Ecol.*, **67**, 893–921.

Thompson, P.A. (1975) Characterization of the germination responses of *Silene dioica* (L.) Clairv. populations from Europe, *Ann. Bot.*, **39**, 1–19.

Tietema, T. (1980) Ecophysiology of the sand sedge *Carex arenaria* L. II. The distribution of 14-C assimilates, *Acta. Bot. Neerl.*, **29**, 165–78.

Tiffney, B. (1984) Seed size, dispersal syndromes, and the rise of the angiosperms: evidence and hypothesis, *Ann. Missouri Bot. Gard.*, **71**, 551–76.

Tilman, D. (1982) *Resource competition and community structure.* Princeton Univ. Press, Princeton, NJ.

Toky, O.P. and Ramakrishnan, P.S. (1981) Cropping and yields in agricultural systems of the north-eastern hill region of India, *Agro-ecosystems*, **7**, 11–25.

Tobey, R.C. (1981) *Saving the prairies: The life cycle of the founding school of American plant ecology, 1895–1955.* Univ. California Press, Berkeley and London.

Tremlett, M. Silvertown, J.W. and Tucker, C. (1984) An analysis of spatial and temporal variation in seedling survival of a monocarpic perennial, *Conium maculatum*, *Oikos*, **43**, 41–5.

Trenbath, B.R. (1974) Biomass productivity of mixtures, *Adv. Agron.*, **26**, 177–210.

Trenbath, B.R. and Harper, J.L. (1973) Neighbour effects in the genus *Avena*: I. Comparison of crop species, *J. Appl. Ecol.*, **10**, 379–400.

Tripathi, R.S. and Harper, J.L. (1973) The comparative biology of *Agropyron repens* L. (Beav.) and *A. caninum* L. (Beav.): 1. The growth of mixed populations established from tillers and from seeds, *J. Ecol.*, **61**, 353–68.

Tubbs, C.R. (1968) *The new forest: an ecological history*. David and Charles, Newton Abbot, Devon.

Turkington, R.A. Cavers, P.B. and Aarssen, L.W. (1977) Neighbour relationships in grass-legume communities: I. Interspecific contacts in four grassland communities near London, Ontario, *Can. J. Bot.*, **55**, 2701–11.

Turkington, R.A. and Harper, J.L. (1979) The growth, distribution and neighbour relationships of *Trifolium repens* in a permanent pasture: 4. Fine-scale biotic differentiation, *Can. J. Bot.*, **57**, 245–54.

Turkington, R. Harper, J.L. de Jong, P. and Aarssen, L.W. (1985) A reanalysis of interspecific association in an old pasture, *J. Ecol.*, **73**, 123–31.

Turnbull, C.L. and Culver, D.C. (1983) The timing of seed dispersal in *Viola nuttallii*: attraction of dispersers and avoidance of predators, *Oecologia*, **59**, 360–5.

Turner, M.D. and Rabinowitz, D. (1983) Factors affecting frequency distributions of plant mass: the absence of dominance and suppression in competing monocultures of *Festuca paradoxa*, *Ecology*, **64**, 469–75.

Uhl, C. and Clark, K. (1983) Seed ecology of selected Amazon basin successional species, *Bot. Gaz.*, **144**, 419–25.

van Baalen, J. (1982) Germination ecology and seed population dynamics of *Digitalis purpurea*, *Oecologia*, **53**, 61–7.

van Baalen, J. and Prins, E.G.M. (1983) Growth and reproduction of *Digitalis purpurea* in different stages of succession, *Oecologia*, **58**, 84–91.

Vance, R.R. (1984) Interference competition and the coexistence of two competitors on a single limiting resource, *Ecology*, **65**, 1349–57.

van Groenendael, J. (1985) Selection for different life histories in *Plantago lanceolata*. Ph.D. Thesis, University of Nijmegen, Netherlands.

van Groenendael, J. and Slim, P. (in press) The contrasting dynamics of two populations of *Plantago lanceolata* L., classified by age and size.

Van Valen, L. (1975) Life, death, and energy of a tree, *Biotropica*, **7**, 260–9.

Varley, G.C. Gradwell, G.R. and Hassell, M.P. (1973) *Insect population ecology*. Blackwell, Oxford; University of California Press, USA.

Vasek, F.C. (1980) Creosote bush: long-lived clones in the Mojave desert, *Am. J. Bot.*, **67**, 246–55.

Vazquez-Yanes, C. and Smith, H. (1982) Phytochrome control of seed germination in the tropical rain forest pioneer trees *Cecropia obtusifolia* and *Piper auritum* and its ecological significance, *New Phytol.*, **92**, 477–86.

Venable, D.L. and Lawlor, L. (1980) Delayed germination and dispersal in desert annuals: escape in space and time, *Oecologia*, **46**, 272–82.

Venable, D.L. and Levin, D.A. (1985) Ecology of achene dimorphism in *Heterotheca latifolia*. I. Achene structure, germination and dispersal, *J. Ecol.*, **73**, 133–45.

Venable, D.L. (1985) The evolutionary ecology of seed heteromorphism. *Am. Nat.*, **126**, 577–95.

Walker, J. and Peet, R.K. (1983) Composition and species-diversity of pine–wiregrass savannas of the Green Swamp, North Carolina, *Vegetatio*, **55**, 163–79.

Waller, D.M. (1981) Neighbourhood competition in several violet populations, *Oecologia*, **51**, 116–22.

Waller, D.M. (1982) Factors influencing seed weight in *Impatiens capensis* (Balsaminaceae), *Am. J. Bot.*, **69**, 1470–75.

Waller, D.M. (1985) The genesis of size hierarchies in seedling populations of *Impatiens capensis* Meerb. *New Phytol.*, **100**, 243–60.

Waloff, N. and Richards, O.W. (1977) The effect of insect fauna on growth, mortality and natality of broom *Sarothamnus scoparius*, *J. Appl. Ecol.*, **14**, 787–98.

Warner, R.R., and Chesson, P.L. (1985) Coexistence mediated by recruitment fluctuations: a field guide to the storage effect, *Am. Nat.*, **125**, 769–87.

Waters, W.E. (1969) The life table approach to an analysis of insect impact, *J. For.*, **67**, 300–4.

Watkinson, A.R. (1978a) The demography of a sand dune annual *Vulpia fasciculata*: II. The dynamics of seed populations, *J. Ecol.*, **66**, 35–44.

Watkinson, A.R. (1978b) The demography of a sand dune annual *Vulpia fasciculata*: III. The dispersal of seeds, *J. Ecol.*, **66**, 483–98.

Watkinson, A.R. (1984) Yield-density relationships: the influence of resource availability on growth and self-thinning in populations of *Vulpia fasciculata*, *Ann. Bot.*, **53**, 469–82.

Watkinson, A.R. (1985) On the abundance of plants along an environmental gradient, *J. Ecol.*, **73**, 569–78.

Watkinson, A.R. and Harper, J.L. (1978) The demography of a sand dune annual *Vulpia fasciculata*: I. The natural regulation of populations, *J. Ecol.*, **66**, 15–33.

Watkinson, A.R. Lonsdale, W.M. and Firbank, L.G. (1983) A neighbourhood approach to self-thinning *Oecologia*, **56**, 381–4.

Watson, M.A. (1979) Age structure and mortality within a group of closely related mosses, *Ecology*, **60**, 988–97.

Watson, M.A. (1984) Developmental constraints: effect on population growth and patterns of resource allocation in a clonal plant, *Am. Nat.*, **123**, 411–26.

Watson, M.A. and Casper, B.B. (1984) Morphogenetic constraints on patterns of carbon distribution in plants, *Ann. Rev. Ecol. Syst.*, **15**, 233–58.

Watt, A.S. (1974) Senescence and rejuvenation in ungrazed chalk grassland (grassland B) in Breckland: the significance of litter and moles, *J. Appl. Ecol.*, **11**, 1157–71.

Watt, A.S. and Fraser, G.K. (1933) Tree roots and the field layer, *J. Ecol.*, **21**, 404–14.

Weaver, S.E. and Cavers, P.B. (1979) The effects of emergence date and emergence order on seedling survival rates in *Rumex crispus* and *R. obtusifolius*, *Can. J. Bot.*, **57**, 730–8.

Weiner, J. (1985) Size hierarchies in experimental populations of plants, *Ecology*, **66**, 743–52.

Weiner, J. and Solbrig, O.T. (1984) The meaning and measurement of size hierarchies in plant populations, *Oecologia*, **61**, 334–6.

Wellbank, P.J. (1963) A comparison of competitive effects of some common weed species, *Ann. Appl. Biol.*, **51**, 107–25.

Welch, D. (1985) Studies in the grazing of heather moorland in North-East Scotland, *J. Appl. Ecol.*, **22**, 461–72.

Werner, P. A. (1975) Predictions of fate from rosette size in teasel (*Dipsacus fullonum* L.), *Oecologia*, **20**, 197–201.

Werner, P.A. (1979) Competition and coexistence of similar species, Ch. 12, pp. 287–310 in Solbrig, O.T., Jain, S. Johnson, G.B. and Raven, P.H. (eds), *Topics in plant population biology*. Macmillan, London and New York.

Werner, P.A. and Caswell, H. (1977) Populations growth rates and age vs. stage-distribution models for teasel (*Dipsacus sylvestris* Huds.), *Ecology*, **58**, 1103–111.

Werner, P.A. and Platt, W.J. (1976) Ecological relationships of co-occurring golden rods (*Solidago*: compositae), *Am. Nat.*, **110**, 959–71.

Wesson, G. and Wareing, P.F. (1969a) The role of light in the germination of naturally occurring populations of buried weed seeds, *J. Exp. Bot.*, **20**, 402–13.

Wesson, G. and Wareing, P.F. (1969b) The induction of light sensitivity in weed seeds by burial, *J. Exp. Bot.*, **20**, 414–25.

West, N.E. Rea, K.H. and Harniss, R.O. (1979) Plant demographic studies in sagebrush-grass communities of southeastern Idaho, *Ecology*, **60**, 376–88.

Westoby, M. (1984) The self-thinning rule, *Adv. Ecol. Res.*, **14**, 167–225.

Whelan, R.J. and Main, A.R. (1979) Insect grazing and post-fire plant succession in south-west Australian woodland, *Aust. J. Ecol.*, **4**, 387–98.

Whipple, S.A. (1978) The relationship of buried, germinating seeds to vegetation in an old growth colorado sub-alpine forest, *Can. J. Bot.*, **56**, 1505–9.

White, J. (1979) The plant as a metapopulation, *Ann. Rev. Ecol. Syst.*, **10**, 109–45.

White, J. (1980) Demographic factors in populations of plants, Ch. 2, pp. 21–48 in Solbrig, O.T. (ed.), *Demography and evolution in plant populations*. Blackwell, Oxford; University of California Press, California.

White, J. (1981) The allometric interpretation of the self-thinning rule. *J. Theor. Biol.*, **89**, 475–500.

White, J. and Harper, J.L. (1970) Correlated changes in plant size and number in plant populations, *J. Ecol.*, **58**, 467–85.

White, P.S. (1979) Pattern, process, and natural disturbance in vegetation, *Bot. Rev.*, **45**, 229–99.

Whitehead, F.H. (1971) Comparative autecology as a guide to plant distribution, pp. 167–76 in Duffey, E.O. and Watt, A.S. (eds), *The scientific management of animal and plant communities for conservation*. 11th Symp. Brit. Ecol. Soc., Blackwell, Oxford; F.A. Davies and Co., Philadelphia, USA.

Whitmore, T.C. (1977) Gaps in the forest canopy, Ch. 27, pp. 639–55 in Tomlinson, P.B. and Zimmerman, M.H. (eds), *Tropical trees as living systems*. Cambridge University Press, Cambridge and New York.

Wiens, D. (1984) Ovule survivorship, brood size, life history, breeding systems, and reproductive success in plants, *Oecologia*, **64**, 47–53.

Wilbur, H.M. (1976) Life history evolution of seven milkweeds of the genus *Asclepias*, *J. Ecol.*, **64**, 223–40.

Wilbur, H.M. (1977) Propagule size number, and dispersion pattern in *Ambystoma* and *Asclepias*, *Am. Nat.*, **111**, 43–68.

Willey, R.W. (1979a) Intercropping: its importance and research needs Part 1. Competition and yield advantage, *Field Crop. Abstr.*, **32**, 1–10.

Willey, R.W. (1979b) Intercropping: its importance and research needs Part 2. Agronomy and research approaches, *Field Crop Abstr.*, **32**, 73–85.

Willey, R.W. and Heath, S.B. (1969) The quantitative relationships between plant population and crop yield, *Adv. Agron.*, **21**, 281–321.

Williams, O.B. (1970) Population dynamics of two perennial grasses in Australian semi-arid grassland, *J. Ecol.*, **58**, 869–75.

Williams, R.D. (1964) Assimilation and translocation in perennial grasses, *Ann. Bot.*, **28**, 419–29.

Williamson, M.H. (1972) *The analysis of biological populations*. Arnold, London.

Wilson, M.F. and Bertin, R.I. (1979) Flower-visitors, nectar production and inflorescence size of *Asclepias syriaca*, *Can. J. Bot.*, **57**, 1380–8.

Willson, M.F. and Burley, N. (1983) *Mate choice in plants*. Monographs in population biology No. 19. Princeton University Press, Princeton, NJ.

Willson, M.F. and Price, P.W. (1977) The evolution of inflorescence size in *Asclepias* (Asclepiadaceae), *Evolution*, **31**, 495–511.

Willson, M.F. and Rathcke, B.J. (1974) Adaptive design of the floral display in *Asclepias syriaca* L., *Am. Mid. Nat.*, **92**, 47–57.

Wit, C.T. de (1960) On competition, *Versl. Landbouwk. Onderz.*, **66**, 1–82.

Wright, H.E. Jr and Heinselman, M.L. (eds) (1973) The ecological role of fire in natural conifer forests of western and northern America, *Quat. Res.*, **3**, 317–513.

Wright, S.J. (1983) The dispersion of eggs by a bruchid beetle among *Scheelea* palm seeds and the effect of distance to the parent palm, *Ecology*, **64**, 1016–21.

Wyatt, R. (1976) Pollination and fruit-set in *Asclepias*: a reappraisal, *Am. J. Bot.*, **63**, 845–51.

Wyatt, R. (1980) The reproductive biology of *Asclepias tuberosa*: I. Flower number, arrangement, and fruit-set, *New Phytol.*, **85**, 119–31.

Wyatt Smith, J. (1958) Seedling/sapling survival of *Shorea leprosula*, *S. parviflora* and *Koompassia malaccensis*, *Malay For.*, **21**, 185–93.

Yakovlev, M.S. and Zhukova, G.Y. (1980) Chlorophyll in embryos of angiosperm seeds, *Bot. Notiser*, **133**, 323–36.

Yarranton, G.A. (1966) A plotless method of sampling vegetation, *J. Ecol.*, **54**, 229–37.

Yeaton, R.I. Yeaton, R.W. Waggoner III, J.P. and Horenstein, J.E. (1985) The ecology of *Yucca* (Agavaceae) over an environmental gradient in the Mohave Desert: distribution and interspecific interactions, *J. Arid. Envts*, **8**, 33–44.

Yoda, K. Kira, T. Ogawa, H. and Hozumi, K. (1963) Self-thinning in overcrowded pure stands under cultivated and natural conditions, *J. Biol. Osaka City Univ.*, **14**, 107–29.

Index

Note: Common names are in parentheses after Latin names, where they are given in the text. They are also entered separately where significant, e.g. buttercups; grasses, etc.